Is it possible that extraterrestrial life forms exist within our Galaxy, the Milky Way? This book offers a critical analysis, by leading experts in a range of sciences, of the plausibility that intelligent life forms do exist. Exploration of the solar system, and observations with telescopes that probe deep space, have come up empty-handed in searches for evidence of extraterrestrial life. Many experts in the fields of astronomy, biology, chemistry and physics now argue that the evidence points to the conclusion that technological civilizations are rare. After ten billion years, and among hundreds of billions of stars, we may well possess the most advanced brains in the Milky Way Galaxy. This second edition elucidates many new aspects of research on extraterrestrial intelligent life, especially biological considerations of the question.

EXTRATERRESTRIALS: WHERE ARE THEY?

EXTRATERRESTRIALS
Where are They?

Edited by

BEN ZUCKERMAN and MICHAEL H. HART

Second Edition

CAMBRIDGE
UNIVERSITY PRESS

Published by the Press Syndicate of the University of Cambridge
The Pitt Building, Trumpington Street, Cambridge CB2 1RP
40 West 20th Street, New York, NY 10011-4211, USA
10 Stamford Road, Oakleigh, Melbourne 3166, Australia

First published by Pergamon Press Inc. 1982
Second edition published by Cambridge University Press 1995

Printed in Great Britain at the University Press, Cambridge

A catalogue record for this book is available from the British Library

Library of Congress cataloguing in publication data

Extraterrestrials – where are they? / edited by Ben Zuckerman and
Michael H. Hart. – 2nd ed.
p.   cm.
Hart's name appears first on the earlier edition.
Includes bibliographical references and index.
ISBN 0-521-44335-0. – ISBN 0-521-44803-4 (pbk)
1. Life on other planets – Congresses. 2. Life – Origin –
Congresses. I. Zuckerman, Ben, 1943–   II. Hart, Michael H.
QB54.E95   1995
574.999 – dc20   94-43739   CIP

ISBN 0 521 44335 0 hardback
ISBN 0 521 44803 4 paperback

Contents

Preface to the Second Edition

The purpose of a second edition of *Where Are They?* is to enlarge upon and update issues that were debated in 1978 at a two-day Symposium of the same name. As might be expected, comparison of the present book with the first edition shows that relatively little has changed in the field of interstellar travel and colonization – there are not many interstellar travelers among us. By way of contrast, because the world is full of biologists and astronomers, there have been many new experiments and insights in these fields. We are especially pleased that three distinguished biologists – Drs Diamond, Joyce and Mayr – have contributed new chapters to the present volume. Typically, biologists appear to be less sanguine about the likelihood of abundant intelligent life in the universe than are engineers and physicists.

At the time of publication of the first edition, the only things that humans knew of with certainty that orbited stars other than the Sun were other stars. Then, in 1983, NASA's IRAS satellite discovered that many nearby stars similar to our Sun emit infrared (heat) radiation well in excess of that expected from their visible surfaces. In a chapter in the first edition entitled 'Searches for electromagnetic signals ... extraterrestrial beings' I discussed the possibility that IRAS might discover a so-called 'Dyson Sphere'. In this picture, initially proposed by Freeman Dyson, an advanced civilization builds and lives in a multitude of space colonies which orbit and absorb light from a central star; these colonies then emit their waste heat into space at infrared wavelengths. Various arguments and follow-up ground-based observations have shown conclusively that the infrared energy discovered by IRAS is not, in general, a signpost of Dyson Spheres but rather of immense numbers of submillimeter-sized particles that orbit the stars. These particles are one of various recent discoveries that, taken together, have convinced all but perhaps the most conservative of astronomers that rocky planets with masses similar to that of the Earth are reasonably commonplace in our Milky Way Galaxy (see chapter by Jon Lunine). So, the answer to the question 'Where are they?' is not 'planetary systems are extremely rare', a possibility that still appeared to be viable when J. P. Harrington wrote his chapter on 'The frequency of planetary systems in the Galaxy' for the first edition.

Nonetheless, even if Earth-mass planets are common in the Galaxy, discoveries and calculation since the time of the first edition still allow that the rarity of technological civilizations might be attributable to peculiarities of solar system formation. Both Jon Lunine and Gerald Joyce consider the idea that frequent impacts of massive bodies onto Earth can have devastating consequences for life, and Jon Lunine discusses why Jupiter-mass planets may be required in all solar systems if Earth-like planets are to be shielded from too frequent bombardment. But, at present, there is neither observational nor theoretical support for the notion that Jupiter-like planets are commonly formed.

At a level even closer to home, the vague but lingering notion that the existence of Earth's large Moon was somehow essential for the origin and/or evolution of Earth-life has recently acquired a much more substantial foothold. The tilt of the rotation axis of the Earth with respect to the perpendicular to the plane of its orbit around the Sun is called the 'obliquity' of the Earth; the reason that we have seasons is because this tilt is not zero. Recently, two different researchers, Laskar & Wisdom (and colleagues), have argued that, in the absence of the stabilizing effect of a large close moon, the obliquity of an Earth-like planet would fluctuate widely and rapidly with concomitant changes in the seasons, devastating life. If, as is probably true, only few Earth-like planets have large moons, then this too could supply an answer to the 'Where are they?' question.

Prospects for future research germane to *Where Are They?* are mixed. The end of the cold war has not translated into increased spending for relevant science and technology; beating the Soviets into space no longer motivates our leaders. Examples of post-cold-war cost cutting are the retreat of the space program and the congressionally mandated termination of NASA's long-term SETI (search for extraterrestrial intelligence) project as the USA struggles to keep pace with economically competitive rising stars, especially in Asia.

Underlying the 'Where are they?' question is the implicit assumption that a technological civilization will ultimately find its way into space, probably first in the form of a Dyson Sphere and then as interstellar voyagers. When might we reasonably hope to take the first step, colonization of our solar system? Unfortunately, because the human race is squandering its natural capital at an unsustainable rate, it seems at least as likely that terrestrial collapse rather than extraterrestrial expansion awaits us in the next two centuries. To borrow a few words from the preface to the first edition: If we do not gain control of our runaway population and rapacious appetites, then the radiation of intelligence throughout the Milky Way will have to wait for a distant time and place.

Ben Zuckerman
Los Angeles, California
April, 1995

Preface to the First Edition

Where are they? Enrico Fermi is reputed to have asked this question at the dawn of the atomic age. He must have been wondering why, having discovered and tamed nuclear energy sources, advanced extraterrestrials were not in evidence here on Earth or out in the skies.

During the 1960s and early 1970s, Fermi's question was largely forgotten or ignored. Advances in radioastronomy, the American and Soviet space programs, the blossoming of the study of molecular biology and progress in laboratory simulations of prebiological organic chemistry all contributed, in their own way, to a euphoric belief among many scientists that life in the cosmos is commonplace and might even be discovered soon. At a more popular level, numerous reports of close encounters of the second and third kind, lavishly bankrolled science fiction movies and enormously popular books on ancient astronauts all served to promote the idea that They are out there and will soon be, or already have been, here.

The past few years have seen the introduction of new and sobering input into this picture. The US program of planetary exploration, while highly successful from a technological and scientific standpoint, has failed to produce even a hint of an extraterrestrial biology. Although the search for simple nonterrestrial life in our solar system cannot be considered complete, the prospects for eventual success do not look good. In addition, searches for evidence of advanced technology, either in deep space or in the solar system, have been discouraging. To astronomers who work with giant optical and radio telescopes the universe appears to be a gigantic wilderness area untouched by the hand of intelligence (with the possible exception of God's). The absence of advanced extraterrestrials on the Earth and, very probably, in the solar system has been interpreted by various scientists, beginning with Michael Hart's 1975 paper that leads off this volume, as evidence that such creatures may not exist anywhere in our Milky Way Galaxy. Thus, Fermi's question has now re-emerged to haunt our minds.

Sometime in 1978 it occurred to Drs Hart, Papagiannis and myself that the scientific community was scarcely aware of many of these ideas and the public at large even less so. So we organized a modest meeting – 'Where Are They? A Symposium on the Implications of Our Failure to Observe Extraterrestrials' – that was held in College Park, Maryland, at the Center of Adult Education of the University of Maryland on 2 and 3 November 1979. The size of the symposium was kept small to encourage and facilitate back and forth discussion among the participants. Some of this discussion, which was tape recorded, is included in the present book following the various papers.

The first day of the conference was devoted to a general overview of the question and the current observational situation, followed by a session on the feasibility of interstellar travel and colonization. On the second day the biological and astronomical contexts of the problem were explored. We believe that many of the papers that were presented at the symposium and in the present volume suggest a significantly different picture from the one that has been painted in most earlier meetings, NASA-supported studies and compendia on the search for extraterrestrial intelligence.

An image of the Milky Way abounding in all kinds of advanced supercivilizations and bizarre forms of life is tremendously appealing to many scientists and lay persons. Then life is but a commonplace extension of cosmic evolution following the Big Bang and we human beings are insignificant – mere cosmic insects. Perhaps this accounts in small part for the indifferent and even ruthless way that many human beings treat life on Earth.

But what if this is not so? What if after 10 billion years and hundreds of billions of stars we possess the most advanced brains in the Galaxy? Surely, if we are alone, then we were meant to play a more noble role than that currently in evidence on our troubled globe. If we are to travel to the stars in the next millennium, then we must, in the next century at most, gain control of our runaway population, rapacious appetites, and nuclear and biological technologies. If we fail to do this, then the radiation of intelligence throughout the Milky Way may have to wait for a distant time and place.

Ben Zuckerman
College Park, Maryland
December, 1980

1
An Explanation for the Absence of Extraterrestrials on Earth*

MICHAEL H. HART

Are there intelligent beings elsewhere in our Galaxy? This is the question which astronomers are most frequently asked by laymen. The question is not a foolish one; indeed, it is perhaps the most significant of all questions in astronomy. In investigating the problem, we must therefore do our best to include all relevant observational data.

Because of their training, most scientists have a tendency to disregard all information which is not the result of measurements. This is, in most matters, a sensible precaution against the intrusion of metaphysical arguments. In the present matter, however, that policy has caused many of us to disregard a clearly empirical fact of great importance, to wit: *There are no intelligent beings from outer space on Earth now*. (There may have been visitors in the past, but none of them have remained to settle or colonize here.) Since frequent reference will be made to the foregoing piece of data, in what follows we shall refer to it as 'Fact A'.

Fact A, like all facts, requires an explanation. Once this is recognized, an argument is suggested which indicates an answer to our original question. If, the argument goes, there were intelligent beings elsewhere in our Galaxy, then they would eventually have achieved space travel, and would have explored and colonized the Galaxy, as we have explored and colonized the Earth. However, (Fact A), they are not here; therefore they do not exist.

The author believes that the above argument is basically correct; however, in the rather loose form stated above it is clearly incomplete. After all, might there not be some other explanation of Fact A? Indeed, many other explanations of Fact A have been proposed: however, none of them appears to be adequate.

The other proposed explanations of Fact A might be grouped as follows:

* Reprinted from the *Quarterly Journal of the Royal Astronomical Society*, **16**, 128–35 (1975), with permission of Blackwell Scientific Publications, Ltd.

1. All explanations which claim that extraterrestrial visitors have never arrived on earth because some physical, astronomical, biological or engineering difficulty makes space travel infeasible. We shall refer to these as 'physical explanations'.
2. Explanations based on the view that extraterrestrials have not arrived on Earth because they have chosen not to. This category is also intended to include any explanation based on their supposed lack of interest, motivation or organization, as well as political explanations. We shall refer to these as 'sociological explanations'.
3. Explanations based on the possibility that advanced civilizations have arisen so recently that, although capable and willing to visit us, they have not had time to reach us yet. We shall call these 'temporal explanations'.
4. Those explanations which take the view that the Earth *has* been visited by extraterrestrials, though we do not observe them here at present.

These four categories are intended to be exhaustive of the plausible alternatives to the explanation we suggest. Therefore, if the reasoning in the next four sections should prove persuasive, it would seem very likely that we are the only intelligent beings in our Galaxy.

Physical Explanations

After the success of Apollo 11 it seems strange to hear people claim that space travel is impossible. Still, the problems involved in interstellar travel are admittedly greater than those involved in a trip to the Moon, so it is reasonable to consider just how serious the problems are, and how they might be overcome.

The most obvious obstacle to interstellar travel is the enormity of the distances between the stars, and the consequently large travel times involved. A brief computation should make the difficulty clear: The greatest speeds which manned aircraft, or even spacecraft, have yet attained is only a few thousand $km\,h^{-1}$. Yet, even travelling at 10% of the speed of light ($\sim$ 1 billion $km\,h^{-1}$), a one-way trip to Sirius, which is one of the nearest stars, would take 88 years. Plainly, the problem presented is not trivial; however, there are several possible means of dealing with it:

1. If it is considered essential that those who start on the voyage should still be reasonably youthful upon arrival, this could be accomplished by having the voyagers spend most of the trip in some form of 'suspended animation'. For example, a suitable combination of drugs might not only put a traveller to sleep, but also slow his metabolism down by a factor of 100 or more. The same result might be effected by freezing the space voyagers near the beginning of the trip, and thawing them out shortly before arrival. It is true that we do not yet know how to freeze and revive warm-blooded animals but: (a) future biologists on Earth (or biologists in

advanced civilizations elsewhere) may learn how to do so; (b) intelligent beings arising in other solar systems are not necessarily warm-blooded.

2. There is no reason to assume that all intelligent extraterrestrials have lifespans similar to ours. (In fact, future medical advances may result in human beings having life expectancies of several millennia, or even perhaps much longer.) For a being with a lifespan of 3000 years a voyage of 200 years might seem not a dreary waste of most of one's life, but rather a diverting interlude.

3. Various highly speculative methods of overcoming the problem have been proposed. For example, utilization of the relativistic time-dilation effect has been suggested (though the difficulties in this approach seem extremely great); or the spaceship might be 'manned' by robots, perhaps with a supplementary population of frozen zygotes which, after arrival at the destination, could be thawed out and used to produce a population of living beings.

4. The most direct manner of handling the problem, and the one which makes the fewest demands on future scientific advances, is the straightforward one of planning each space voyage, from the beginning, as one that will take more than one generation to complete. If the spaceship is large and comfortable, and the social structure and arrangements are planned carefully, there is no reason why this need be impracticable.

Another frequently mentioned obstacle to interstellar travel is the magnitude of the energy requirements. This problem might be insurmountable if only chemical fuels were available; but if nuclear energy is used, the fuel requirements do not appear to be extreme. For example, the kinetic energy of a spaceship travelling at one-tenth the speed of light is:

$$KE = (\gamma-1)\,Mc^2 = ([1.0-0.01]^{-1/2}-1)\,Mc^2 = 0.005\,Mc^2 \qquad (1.1)$$

Now the energy released in the fusion of a mass F of hydrogen into helium is approximately $0.007\,Fc^2$. In principle, the mechanical efficiency of a nuclear-powered rocket can be more than 60% (Von Hoerner, 1962; Marx, 1963). However, let us assume that in practice only one-third of the nuclear energy could actually be released and converted into kinetic energy of the spacecraft. Then the fuel needed to accelerate the spaceship to $0.10\,c$ is given by:

$$0.005\,Mc^2 = 0.007\,Fc^2/3 \qquad (1.2)$$

This gives: $F = 2.14\,M$ and $T = 3.14\,M$, where T is the combined mass of spaceship and fuel. The necessity of starting out with enough fuel first to accelerate the ship, and later to decelerate, introduces another factor of 3.14: so initially we must have $T = 9.88\,M$. In other words, the ship must start its voyage carrying about nine times its own weight in fuel. This is a rather modest requirement, particularly in view of the cheapness and abundance of the fuel. (The enormous fuel-to-payload ratios computed by Purcell (1963) are a result of his considering only relativistic space flight; a travel speed of $0.1\,c$ seems more realistic.) Furthermore, there are several possible ways of reducing the fuel-to-payload ratio, including (a) refuelling from auxiliary craft; (b) scooping up H atoms while traveling through interstellar space; (c) greater engine efficiencies; (d) traveling at slightly lower speeds (traveling at 0.09 c instead of 0.10 c would reduce the fuel-to-payload ratio to 6.5:1); and (e) using methods of propulsion other than rockets. For some interesting possibilities, see Marx (1966) and other papers listed by Mallove & Forward (1972).

It can be seen that neither the time of travel nor the energy requirements create an insuperable obstacle to space travel. However, in the past, it was sometimes suggested that one or more of the following would make space travel unreasonably hazardous: (a) the effects of cosmic rays; (b) the danger of collisions with meteoroids; (c) the biological effects of prolonged weightlessness; and (d) unpredictable or unspecified dangers. With the success of the Apollo and Skylab missions it appears that none of these hazards is so great as to prohibit space travel.

Sociological Explanations

Most proposed explanations of Fact A fall into this category. A few typical examples are:

(1) Why take the anthropomorphic view that extraterrestrials are just like us? Perhaps most advanced civilizations are primarily concerned with spiritual contemplation and have no interest in space exploration (the Contemplation Hypothesis.)

(2) Perhaps most technologically advanced species destroy themselves in nuclear warfare not long after they discover atomic energy (the Self-destruction Hypothesis.)

(3) Perhaps an advanced civilization has set the Earth aside as their version of a national forest, or wildlife preserve (the Zoo Hypothesis: Ball, 1973).

In addition to variations on these themes (for example, extraterrestrials might be primarily concerned with artistic values rather than spiritual contemplation), many quite different explanations have been suggested. Plainly, it is not possible to consider each of these individually. There is, however, a weak spot which is common to all of these theories.

Consider, for example, the Contemplation Hypothesis. This might be a perfectly adequate explanation of why, in the year 600 000 BC, the inhabitants of Vega III chose not to visit the Earth. However, as we well know, civilizations and cultures change. The Vegans of 599 000 BC could well be less interested in spiritual matters than their ancestors were, and more interested in space travel. A similar possibility would exist in 598 000 BC, and so forth. Even if we assume that the Vegans' social and political structure is so rigid that no changes occur even over hundreds of thousands of years, or that their basic psychological makeup is such that they always remain uninterested in space travel, there is still a problem.

With such an additional assumption the Contemplation Hypothesis might explain why the Vegans have never visited the Earth, but it still would not explain why the civilizations which developed on Procyon VI, Sirius II and Altair IV have also failed to come here. The Contemplation Hypothesis is not sufficient to explain Fact A unless we assume that it will hold for *every* race of extraterrestrials – regardless of its biological, psychological, social or political structure – and at every stage in their history after they achieve the ability to engage in space travel. That assumption is not plausible, however, so the Contemplation Hypothesis must be rejected as insufficient.

The same objection, however, applies to any other proposed sociological explanation. No such hypothesis is sufficient to explain Fact A unless we can show that it will apply to every race in the Galaxy, and at every time.

The foregoing objection would hold even if there *were* some established sociological theory which predicted that most technologically advanced civilizations will be spiritually oriented, or will blow themselves up, or will refrain from exploring and colonizing. In point of fact, however, there is no such theory which has been generally accepted by political scientists, or sociologists, or psychologists. Furthermore, it is safe to say that no such theory will be accepted, for any scientific theory must be based upon evidence, and the only evidence concerning the behaviour of technologically advanced civilizations which political scientists, sociologists and psychologists have comes from the human species – a species which has neither blown itself up nor confined itself exclusively to spiritual contemplation, but which has explored and colonized every portion of the globe it could. (This is *not* intended as proof that all extraterrestrials must behave as we have; it *is* intended to show that we cannot expect a scientific theory to be developed which predicts that most extraterrestrials will behave in the reverse way.)

Another objection to any sociological explanation of Fact A is methodological. Faced with a clear physical fact, astronomers should attempt to find a scientific explanation for it – one based on known physical laws and subject to observational or experimental tests. No scientific procedure has ever been suggested for testing the validity of the Zoo Hypothesis, the Self-destruction Hypothesis or any other suggested sociological explanation of Fact A; therefore, to accept any such explanation would be to abandon our scientific approach to the question.

Temporal Explanations

Even if one rejects the physical and sociological explanations of Fact A, the possibility exists that the reason no extraterrestrials are here is simply because none have yet had the time to reach us. To judge how plausible this explanation is, one needs some estimate of how long it might take a civilization to reach us once it had embarked upon a programme of space exploration. To obtain such an estimate, let us reverse the question and ask how long it will be, assuming that we are indeed the first species in our Galaxy to achieve interstellar travel, before we visit a given planet in the Galaxy?

Assume that we eventually send expeditions to each of the 100 nearest stars. (These are all within 20 light-years of the Sun.) Each of these colonies has the potential of eventually sending out their own expeditions, and their colonies in turn can colonize, and so forth. If there were no pause between trips, the frontier of space exploration would then lie roughly on the surface of a sphere whose radius was increasing at a speed of 0.10 c. At that rate, most of our Galaxy would be traversed within 650 000 years. If we assume that the time between voyages is of the same order as the length of a single voyage, then the time needed to span the Galaxy will be roughly doubled.

We see that if there were other advanced civilizations in our Galaxy, they would have had ample time to reach us, unless they commenced space exploration less than 2 million years ago. (There is no real chance of the Sun being accidentally overlooked. Even if the residents of one nearby planetary system ignored us, within a few thousand years an expedition from one of their colonies, or from some other nearby planetary system, would visit the solar system.)

Now the age of our Galaxy is $\sim 10^{10}$ years. To accept the temporal explanation of Fact A, we must therefore hypothesize that (a) it took roughly 5000 time-units (choosing one time-unit $= 2 \times 10^6$ years) for the first species to arise in our Galaxy which had the inclination and ability to engage in interstellar travel; but (b) the second such species (i.e. us) arose less than one time-unit later.

Plainly, this would involve a quite remarkable coincidence. We conclude that, though the temporal explanation is theoretically possible, it should be considered highly unlikely.

Perhaps They Have Come

There are several versions of this theory. Perhaps the most common one is the hypothesis that visitors from space arrived here in the fairly recent past (within, say, the last 5000 years) but did not settle here permanently. There are various interesting archaeological finds which proponents of this hypothesis often suggest are relics of the aliens' visit to Earth.

The weak spot of that hypothesis is that it fails to explain why the Earth was not visited earlier:

1. If it is assumed that extraterrestrials have been able to visit us for a long time, then a sociological theory is required to explain why they all postponed the voyage to Earth for so long. However, any such sociological explanation runs into the difficulties described earlier.

2. On the other hand, suppose it is assumed that extraterrestrials visited us as soon as they were able to. That this occurred within 5000 years (which is only 1/400 of a time-unit) of the advent of our own space age would involve an even more remarkable coincidence than that discussed in the previous section.

Another version of the theory is that the Earth was visited from space a very long time ago, say 50 million years ago. This version involves no temporal coincidence. However, once again, a sociological theory is required to explain why, in all the intervening years, no other extraterrestrials have chosen to come to Earth, and remain. Of course, any suggested mechanism which is effective only 50% (or even 90%) of the time would be insufficient to explain Fact A. (For example, the hypothesis that *most* extraterrestrials wished only to visit, but not to colonize, is inadequate. For colonization not to have occurred requires that *every* single civilization which had the opportunity to colonize chose not to.)

A third version, which we may call 'the UFO Hypothesis', is that extraterrestrials have not only arrived on Earth, but are still here. This version is not really an explanation of Fact A, but rather a denial of it. Since very few astronomers believe the UFO Hypothesis, it seems unnecessary to discuss my own reasons for rejecting it.

Conclusions and Discussion

In recent years several astronomers have suggested that intelligent life in our Galaxy is very common. It has been argued (Shklovskii & Sagan, 1968) that (a) a high percentage of stars have planetary systems; (b) most of these systems contain an Earth-like planet; (c) life has developed on most of such planets; and (d) intelligent life has evolved on a considerable number of such planets. These optimistic conclusions have perhaps led many persons to believe that (1) our starfaring descendants are almost certain, sooner or later, to encounter other advanced cultures in our Galaxy; and (2) radio contact with other civilizations may be just around the corner.

These are very exciting prospects indeed; so much so, that wishful thinking may lead us to overestimate the chances that the conjecture is correct. Unfortunately, though, the idea that thousands of advanced civilizations are scattered throughout the Galaxy is quite implausible in the light of Fact A. Though it is possible that one or two civilizations have evolved and have destroyed themselves in a nuclear war, it is implausible that every one of 10 000 alien civilizations has done so. Our descendants might eventually encounter a few advanced civilizations which never chose to engage in interstellar travel; but their number should be small, and could well be zero.

If the basic thesis of this chapter is correct, there are two corollary conclusions: (1) an extensive search for radio messages from other civilizations is probably a waste of time and money; and (2) in the long run, cultures descended directly from ours will probably occupy most of the habitable planets in our Galaxy.

In view of the enormous number of stars in our Galaxy, the conclusions reached in this chapter may be rather surprising. It is natural to inquire how it has come about that intelligent life has evolved on Earth in advance of its appearance on other planets. Future research in such fields as biochemistry; the dynamics of planetary formation; and the formation and evolution of atmospheres, may well provide a convincing answer to this question. In the meantime, Fact A provides strong evidence that we are the first civilization in our Galaxy, even though the cause of our priority is not yet known.

References

BALL, JOHN A. (1973). *Icarus*, **19**, 347–9.

MALLOVE, E. F. & FORWARD, R. L. (1972). *Bibliography of Interstellar Travel and Communication*, pp. 16–21. Reseach Report 460, Hughes Research Laboratories, Malibu, California.

MARX, G. (1963). *Astronautica Acta*, **9**, 131–9.

MARX, G. (1966). *Nature*, **211**, 22–3.

PURCELL, E. (1963). In *Interstellar Communication*, ed. A. G. W. Cameron, Chapter 13. New York: Benjamin.

SHKLOVSKII, I. S. & SAGAN, C.(1968). *Intelligent Life in the Universe*, Chapter 29. San Francisco: Holden-Day.

VON HOERNER, S. (1962). *Science*, **137**, 18–23.

2
One Attempt To Find Where They Are: NASA's High Resolution Microwave Survey

JILL TARTER

Introduction

Cocconi and Morrison (1959) closed their seminal paper on SETI with a statement that still well characterizes our current situation: 'The probability of success is difficult to estimate, but if we never search the chance of success is zero.' This chapter is a brief summary of how and why NASA has shaped the High Resolution Microwave Survey (HRMS), which it inaugurated on 12 October 1992. Some of the alternative search strategies that were considered are also noted, since these may well form the basis for the next generation of searches, should the HRMS fail to detect a signal.

Although this endeavor is often referred to as SETI (the Search for Extraterrestrial Intelligence), as it is implemented today, and into the foreseeable future, individual search projects are actually seeking evidence of extraterrestrial *technology*. Thus for scientists and engineers engaged in this exploration, a species' *ability to technologically modify its local environment in ways that can be detected over interstellar distances* has become a pragmatic substitute for the overly complex and convoluted definitions of 'intelligence' offered by researchers in other fields. Far in the future lies the promise of being able to detect indirect, but compelling, evidence of life itself on a distant planet. The coexistence of highly reactive gases (such as methane and oxygen) in the atmosphere of a planet, orbiting at an appropriate distance from its host star (so that liquid surface water might be possible) would suggest a continuous biological source at the base of that atmosphere. This chemical disequilibrium would not tell us whether the planet harbored only blue-green algae or creatures like ourselves, but we would at least have a better handle on the 'life' term of the Drake Equation (Drake, 1961). Until our technology is up to this life search, we shall have to be content to search for the manifestations of extraterrestrial technologies. If our own example is representative, it is unlikely that thoughts and intelligence can project themselves across light-years. If, as we suspect, life is a planetary phenomenon, then some day, having detected a distant technology and inferring intelligent life as its creator, we may also have indirectly detected an extrasolar planetary system. Although there are a number of ongoing searches for extrasolar planetary systems, some of which are poised to make significant statements about the existence of jovian-mass planets around nearby stars by the end of this decade, there is a chance that HRMS may make the first serendipitous discovery of planets elsewhere.

Manifestations of Technology

Life, as we know it on Earth, utilizes technology for purposes of providing shelter, producing food, prolonging life, amusement, energy production, transportation, waging war, and communication and information management. It is possible that large-scale astroengineering projects undertaken for most of these purposes could alter the spectral signature of the host star in ways that could be detected. Such alteration is easiest to predict in the case of energy production and has been searched for on several occasions: enhancement of rare earth elements such as praseodymium as the result of the dumping of fissile waste into the star (Freitas, 1985), the 1516 MHz line of tritium (half-life of 12.5 years) in the vicinity of a star could indicate leakage from orbital fusion reactors (Valdes & Freitas, 1984), an infrared excess resulting from a Dyson Sphere capturing and reradiating stellar energy for power (Dyson, 1960). Interstellar transportation by an advanced technology might be detectable as gamma-ray bursts from annihilation rockets (Harris, 1990). Alternatively, space travel might be detected in the form of a companion craft to asteroids within our own solar system, if the voyagers have made a stop and mining operations are in progress (Papagiannis, 1983), or by reflected radiation off patient probes in station-keeping orbits at the Lagrange points of the Sun–Earth–Moon system (Freitas & Valdes, 1980). Planet-destroying warfare might go unnoticed in the glare of the adjacent star, and no intentional search programs for this type of evidence have been reported in the literature.

The problem with all of the examples cited so far is that we do not as yet engage in any of these activities, and thus have no historic analog for which to search. A negative search result does not necessarily mean that the postulated activity is not taking place: it may simply mean that we do not have the required sensitivity to detect it at the level of occurrence. Only in the case of communication and information management can we use ourselves as a metric. We can define what it would take to detect an analog of the twentieth century Earth if it were located anywhere in the Milky Way Galaxy. Any given search may or may not achieve the required sensitivity, but we at least know what we would have to do to make a sufficient search for that particular hypothesis. Another technology, if it exists, will likely be older and presumably more advanced than our own; the probability of its being as primitive or younger than Earth's technology is small. There is considerable debate as to whether their advanced age means that the technologies involved in communication and information management will be more or less detectable than Earth today. With our example of one, this is not a debate that can be settled. However, in the one case of purposefully transmitted signals, it is possible to formulate a testable hypothesis and conduct the experiment. A positive result would answer the age-old question of whether we are alone in this universe: a negative result only constrains one particular hypothesis, the degree of constraint depending on the achieved sensitivity.

What Signals?

An advanced civilization may make use of technologies that we cannot even begin to imagine for the purpose of its own planetary communications and information management. However, once the distances involved reach to interstellar scales, the properties of the interstellar medium begin to restrict what is possible. Certainly our understanding of physics is as yet incomplete, and the unimaginable may be operative over these long scales too, but there is nothing we can do about that except survive to grow wiser. Selection of a search strategy must necessarily be based on what we know, and a choice among several feasible options made on the basis of any economies of resources we think might be universal. The ideal carrier of a bit of information over interstellar distances should:

1. be economical;
2. travel as fast as possible;
3. be easy to generate and receive; and
4. go where it is sent, with minimal absorption or deflection.

It is difficult to assess the economics of an advanced technology, but it is not unreasonable to assume that energy will always have a cost, however it is produced. For an advanced technology, this might not be the limiting resource, but sufficient low-cost energy should make available many other types of resources. With this reasoning, economics suggest that the energy per quantum of information be minimized. To travel at the speed of light, the carrier should be massless.

Instantaneous information transfer by action at a distance still exceeds our grasp and appears impossible, although it lurks tantalizingly around the edges of our physics. Ease of generation and reception rules out exotic massless, or near-massless particles, leaving the photon as a viable candidate. The presence of galactic magnetic fields will deflect charged particles, and interstellar dust and gas will absorb higher-energy photons. This line of argument suggests the selection of low-energy photons, but it is not absolute.

This preference for low-energy photons, combined with a minimum, at radio wavelengths, in the radiated power from natural emission processes on the sky (starlight, infrared cirrus, 3K background, etc., etc.) explains why the majority of searches for signals from other technologies have been, and continue to be conducted with radio telescopes. In space, this quiet window on the cosmos extends from about 1 GHz (lower frequencies become noisy because of galactic synchrotron emission) to 70 or 80 GHz (higher frequencies suffer from quantum noise in detectors as well as increased emission from molecular clouds). Ground-based observations must be conducted through a narrower microwave window limited above 10 GHz by additional atmospheric noise from molecular oxygen and water vapor. This terrestrial microwave window has been the focus of researchers' efforts for over thirty years (Tarter, 1992).

Nature may be quiet in the microwave region of the spectrum, but certainly not silent. What, then, constitutes a signal that is likely to be of technological rather than astrophysical origin? Natural emission tends to be broadband, the narrowest known features from OH masers occupying in excess of 300 Hz of spectrum. Fractional bandwidths can be as small as 10^{-6} for water masers, but we see nothing more coherent, and expect that thermal motions and the Central Limit Theorem will broaden all atomic or molecular emissions. In contrast, our technology is capable of producing microwave signals with fractional bandwidths as small as 10^{-12}. Such signals produce a very large signal-to-noise ratio for a minimum transmitted power. Narrowband signals, either continuous wave or bandlimited pulses, are sought as indicators of distant technologies.

Needles and Haystacks

Looking for a narrowband signal somewhere in the microwave region of the spectrum is often described as looking for a needle in the cosmic haystack. This description is intended to convey how difficult the task is, but it is woefully inadequate. Just how inadequate can be demonstrated after one has spent an afternoon with the local university reference librarian researching statistics on terrestrial hay production. Suppose you were convinced that somewhere on Earth there was a needle of extraordinary value buried in a haystack. If you were willing to spend your entire professional life searching for it, how long would you have to look at each haystack? Assume the needle exists at the beginning of the search and you have 50 years to find it. First, how many haystacks are there? The solid Earth surface is 370 billion acres, of which 1.8 billion acres is potentially arable. Of that, 22% or 390 million acres is under cultivation and 5% of that or 19.5 million acres is devoted to hay. How many haystacks are produced by each acre? The average yield is 20 tons per acre. If each haystack is a cone with a base 6 feet in diameter and a height of 5 feet and if hay has a density of 0.5 grams per cubic centimeter, then each acre produces 20 haystacks. Globally there are about 400 million haystacks. This means that you have about 4 seconds to examine each haystack, assuming that you could travel between them instantaneously! Of course, unless you can manage to store all this hay for 50 years, it would normally be consumed by cows and other hungry creatures in about 6 months. Thus, the time per haystack should more reasonably be set at 0.04 seconds. What does this tell us?

In this exercise, I have guessed at the density of hay and assumed instant transport, so the numbers should not be taken literally. The interesting thing to note is that the number is of the order of seconds, not nanoseconds or femtoseconds. While one person cannot complete this search, many people with existing technology could. That is not the case when it comes to searching for signals from extraterrestrial technologies. With current radioastronomical technologies, it would take an individual billions of years to complete a single search. The cosmic haystack is much bigger than Farmer Jones' haystack!

Having started in this direction, there are other lessons that can be learned from the needle-in-haystack analogy, all of which have their counterparts in SETI. To find the terrestrial needle, you could call up your friends around the world (the cosmic needle needs telescopes worldwide to make a systematic search). You could search all haystacks for a short time (sky survey) or selected haystacks for a longer time (targeted search). You could search the haystacks for many different kinds of needles (i.e. sewing, knitting, steel, plastic, bone) or you could make a highly optimized search for one particular definition of a needle ('magic' frequency, time or place strategies). You could decide to conduct your haystack searches only while the haystacks were being constructed or taken apart for food (commensal searches during astronomical observations). Instead of searching the haystacks you could examine cows, which might yield magnetic needles. If, as is often the case in Europe, they carry magnets in their stomachs, the cows would serve as a concentrator and provide a history of many harvests and many haystacks (searching external galaxies for superstrong signals, thus observing many stars simultaneously is the analogous SETI strategy). The last lesson is that for most types of needles, the hands, eyes and pitchfork are not optimal: better tools are needed (digital signal processors optimized for narrowband signal detection, rather than radioastronomy detectors). As noted, these lessons have all been applied to the cosmic search at one time or another – what are we doing today?

HRMS: Initial Deployments

The United States is by far the dominant country today with respect to active search programs. A summary of SETI activity worldwide, including the US, was presented at the 1992 World Space Congress (Backus, 1995). Here we shall describe the inauguration of NASA's HRMS, which took place shortly after that congress.

HRMS is a bimodal search project. A Sky Survey utilizes the new 34 m beam-waveguide antennas of NASA's Deep Space Network (DSN) of tracking stations to cover the entire terrestrial microwave window (1 – 10 GHz) with modest sensitivity. A Targeted Search uses large radioastronomical telescopes around the world to selectively target solar-type stars. The Targeted Search sacrifices frequency coverage on these large dishes (only 1 – 3 GHz) in order to gain sensitivity for a wider class of continuous and pulsed signals, drifting and/or stationary in frequency.

Using the X-band (8 GHz) receivers on the 34 m antenna at DSS13 within the Goldstone Complex in California's Mojave Desert, the Sky Survey began on 12 October 1992 with a prototype system that has the functional capability of 1/16 of what will be the final operational system. The antenna is driven across the sky in a racetrack pattern at a speed of 0.2 degrees per second. Each block of sky or skyframe defined by this racetrack is 30 degrees long and a few degrees high (depends on frequency) and takes approximately 1.5 hours to scan. A wideband spectrum analyzer (WBSA) functions as a 40 MHz single polarization, or 20 MHz dual-polarization Fourier Transform spectrometer. It uses two million spectral channels whose spacing is 19 Hz to analyze the incoming data from the rapidly moving sky. These spectral data are convolved with a filter representing the beam profile of the antenna, thresholded against an averaged spectrum, and searched for events exceeding a threshold in an online pattern detector called a BECAT (baseline estimation, convolution and thresholding). Events exceeding threshold along a scan line are called singlets, while events that show up on two adjacent scan lines, separated by one half power beamwidth (HPBW), are logged as doublets and represent good candidates for signals that are fixed on the sidereal sky.

The detection software of the Sky Survey makes use of the rapid movement of the antenna to help discriminate against manmade radiofrequency interference (RFI). During the first scan line, signals detected over multiple beams generate masks for RFI for the rest of that skyframe, and signals at a particular frequency that persist over many scan lines are also masked out. Approximately 10 minutes elapse between observations of the same portion of the sky, but on adjacent scan lines. A source of RFI that is fixed to the Earth or in orbit will appear to move relative to the distant stars during that time, and will not produce doublet candidates. Immediately following completion of a skyframe, additional masks are applied to eliminate rapidly drifting interference (such as radiosondes) and the strongest 10 singlet and 10 doublet detections are reobserved in a targeted manner. The telescope executes a conical scan around the detected position to better locate any candidate signal that persists.

When fully operational, the Sky Survey will operate on one of the 34 m antennas in the DSN for 16 hours a day for 7 years in order to complete 30 complete maps of the sky, each 300 MHz wide (thus covering 9 GHz). The Sky Survey has been operating at Goldstone since the inaugural event and has completed 50 X-band skyframes, with a much lower duty cycle on the telescope, as appropriate for the prototype. The observations are now routinely conducted remotely from the Jet Propulsion Laboratory, much as they will be during the fully operational phase. In addition to the X-band observations, the Sky Survey also made use of time available before the decommissioning of the 26 m antenna at DSS 13 to conduct multiple observations of three locations in the Galactic Plane at L-band (1.6 GHz). In all, 150 skyframes were observed, fully sampling these selected regions of the plane with scan line separations equal to ½ HPBW. These data are being used to develop techniques for creating radioastronomical maps of the galactic plane during the course of the Sky Survey. To date only two singlet/doublet candidates have persisted through targeted re-observation, and further observations subsequently discarded these signals. While the RFI experienced at L-band was intermittently extreme, the X-band environment has been benign, permitting the theoretical false alarm probability of 10^{-11} to be routinely achieved. Because the beam-waveguide antennas do not perform well at the very lowest frequencies of the Sky Survey, the Owens Valley Radio Observatory (OVRO) 40 m antenna will be used to conduct that portion of the survey. For 5 months during the summers of 1994 and 1995, the prototype system will be moved to OVRO and used with new wideband RF prime focus feeds and receivers being developed for both the Sky Survey and Targeted Search. Subsequently, cassegrain versions of these systems will be installed on the 34 m antenna for use with the fully operational Sky Survey System.

At the same time that the Sky Survey commenced its observations of a part of the sky visible from Southern California, the Targeted Search began observing

RGO 9548A, one of the stars in that same skyframe, from the Arecibo Observatory in Puerto Rico. This 1000 foot dish is the largest in the world and provided access to some 300 MHz of the spectrum from 1300 to 2300 MHz via a series of line feeds and receivers. The Targeted Search performed 200 hours of observation at Arecibo from 12 October to 15 November 1992 before returning the observing equipment to Ames Research Center, which serves as the Lead Center for the HRMS with a focus on the Targeted Search. The equipment for the initial observations of the Targeted Search was equivalent to what was expected to be ¼ of the fully operational system. It consisted of a dual polarization, IF conversion system and A/D converter that mated with the existing line feeds at the site. A multichannel spectrum analyzer (MCSA), based on a custom VLSI digital signal processing chip, performed Fourier Transforms to provide three simultaneous resolutions with channel bandwidths of 1, 7 and 28 Hz. The data rate for each resolution was the same, with 8 MHz of incoming bandwidth being sorted into 14 million channels at the finest resolution. The outputs from the MCSA were processed in near-real-time by two hardware pattern detectors, a continuous wave and a pulse detector. In both cases, signals which changed their frequency by as much as 1 Hz per second were sought, to allow for some uncompensated relative acceleration between transmitter and receiver. The outputs from the hardware detectors were reported to a control computer, where they were clustered together to give the best representation of the signal (since the detectors were optimized for weak signals, a strong signal produces many different reports of itself) and that was compared against a database of previously seen signals to filter out RFI.

Twenty-five solar-type stars in 24 systems were the targets for these initial observations. In an attempt to understand the temporal structure of the RFI, each target observation was made as a sequence of pointings: On, Off and On again, each pointing lasting 300 or 100 seconds. Only signals that persisted from the first to the second On pointing, but were absent during the Off pointing and had not previously been seen from the direction of some other target, were considered as viable candidates. In all, some 436 On–Off–On observations of the stars at various frequencies were completed and on 15 occasions candidate signals were identified. In subsequent observations these signals were all rejected, but the large number of cases demonstrates the complex temporal nature of RFI, at least on the island of Puerto Rico. Although all of this processing was under software control, the rules for how best to characterize, cluster and match signal reports were tentative, and humans actually monitored the reports from the computer and made (sometimes erroneously) the decision on whether additional observations were warranted. Had we humans followed the suggestions of the computer, a few of the 15 interesting candidates would have been thrown out on the basis of previous observations. On the other hand, the initial rules for matching were far too stringent and the human observer readily recognized patterns that indicated the same signal at a substantially different frequency, presumably due to inexpensive commercial oscillators that drift. The time interval between the end of the Arecibo observations and the present has been spent in trying to create the rule set that will allow software and humanware to agree on the classification and matching of signals, and in turning the lessons learned into innovative new hardware to deal with temporally varying RFI.

Instead of replicating the Arecibo hardware three more times, two more replicas will be made, and the remaining resources will be used to construct follow-up detection devices (FUDD). These take advantage of the fact that the full bandwidth Fourier Transform device is not needed to ascertain whether a previously detected signal is still present as the antenna continues to track the star (although some additional flexible LO systems are needed to select just the right portion of the RF system), and the fact that two widely separated antennas both simultaneously observing the same target can be operated as a virtual interferometer if a noise-free version of the signal detected at one antenna is correlated with the data from the second (and always smaller) antenna. Future Targeted Search observations, covering about 1000 solar-type stars, will always involve two antennas separated by at least several hundred kilometers: the primary Targeted Search system will be on the main (larger) antenna and FUDDs will be on both main and remote antennas. It is hoped that this new strategy will effectively combat the ever-increasing RFI at L-band frequencies. The next observations will take place in Australia in 1994–5, using the Parkes Observatory and the remote Mopra site. When the Gregorian upgrade to Arecibo is completed, additional observing campaigns totalling 2600 hours of time will be undertaken, with the 140 foot telescope at the National Radio Astronomy Observatory in Green Bank, West Virginia serving as the remote antenna. Whenever the new 100 m Green Bank Telescope (GBT) is commissioned, the 140 foot antenna will become a dedicated facility for the Targeted Search, allowing northern hemisphere targets to be observed for 1000 seconds or more to search for long repetition pulses. Either the GBT or Nancay in France will provide the large collecting area needed to complete the Targeted Search of stars too far north for Arecibo.

NASA Pulls Out

This chapter was written prior to 22 September 1993. On that date, Senator Bryan from the state of Nevada introduced a successful amendment to the 1994 appropriations bill that funds NASA, which eliminated funding for HRMS. The conference committee between the legislators of the House and Senate upheld this termination on 1 October. By Congressional mandate, NASA is now out of the SETI business for the foreseeable future. The other US SETI activities in Backus (1995) continue, although each is affected to some degree because NASA had been providing at least partial support for all of them. This does not, however, mean that the sorts of systematic exploration intended for HRMS will not be possible. The taxpayers of the United States have supported this effort for nearly two decades and have invested almost 60 million dollars in special-purpose digital signal processing equipment and search strategies. At this time, the non-profit SETI Institute of Mountain View, California, is actively engaged in seeking private funding to ensure that it will be possible to capitalize on that long investment. For nearly a decade, the SETI Institute has worked in collaboration with NASA to design and implement the specialized equipment for the Targeted Search. The first funds being sought would keep the engineering and science team together and permit the Targeted Search instrumentation and the search of the southern hemisphere from Australia to be carried to completion as planned. The instrumentation would then be installed at Arecibo Observatory to take advantage of the telescope time already awarded there. After that, ways and means to accomplish a systematic survey of the sky without dependence upon the DSN antennas of NASA will be explored, as well as the best means for completing the Targeted Search of the northern hemisphere not visible from Arecibo.

In the one month since Congressional action precipitated a crisis for members of the HRMS team, significant private donations were forthcoming, and there is a sense of optimism at the SETI Institute. It is hoped that a stable, long-term, endowed source of funding can be secured from visionary individuals who wish to see humanity continue to try to satisfy its curiosity about its own place in the universe.

Conclusion

This is a unique time in human history. Some cultures possess the technology and desire to include scientists and engineers in the long-standing debate previously dominated by priests and philosophers and to support them to attempt to end the debate experimentally. This is a beginning whose end cannot be predicted. For those inclined to search, there is at least the hope of a successful conclusion in the foreseeable future. Those who are more pessimistic and disinclined toward searching have effectively answered the question for themselves; for all practical purposes they are alone. How much searching will be done, for how long, will eventually be decided by those who pay the bills. This author is one who would favor an expenditure of time and resources that is consistent with the importance of a question that has pervaded all cultures at least since the beginning of recorded history.

Acknowledgements

The work reported here is the result of the efforts of many scientists and engineers at Ames Research Center and the Jet Propulsion Laboratory. In addition, grants, co-operative agreements and contracts have enabled additional support from an Investigators Working Group of university scientists, the SETI Institute, Silicon Engines, Inc., John Reykjalin, Inc., Sterling Federal Systems, Sverdrup Technology and Hill Sales.

References

BACKUS, P. R. (guest ed.) (1995). SETI-4: SETI, A new endeavor for mankind. Special issue *Acta Astronautica*, in preparation.

COCCONI, G. & MORRISON, P. (1959). Searching for interstellar communications. *Nature*, **184**, 844–6.

DRAKE, F. D. (1961). Project Ozma. *Physics Today*, **14**, 40–2, 44, 46.

DYSON, F. J. (1960). Search for artificial stellar sources of infrared radiation. *Science*, **131**, 1667–8.

FREITAS, R. A. (1985). 21 cm Radio emissions with geometric fine structure (Gray, Dixon, Ehman & Talent). Cited in *JBIS*, **38**, 106.

FREITAS, R. A. & VALDES, F. (1980). A search for natural or artificial objects located at the Earth–Moon libration point. *Icarus*, **42**, 442–7.

HARRIS, M. J. (1990). A search for linear alignments of gamma ray burst sources. *JBIS*, **43**, 551.

PAPAGIANNIS, M. D. (1983). The importance of exploring the asteroid belt. *Acta Astronautica*. **10**, No. 10, 709–12.

TARTER, J. C. (1992). The search for life beyond the solar system. In *Frontiers of Life*, ed. J. & K. Trân Thanh Vân, J. C. Mounolou, J. Schneider & C. McKay, pp. 351–97. Editions Frontières.

VALDES, F. & FREITAS, R. A. (1984). A search for the tritium hyperfine line from nearby stars. Presented at the *35th IAF Congress*, Lausanne, Switzerland.

3

An Examination of Claims That Extraterrestrial Visitors to Earth Are Being Observed

ROBERT SHEAFFER

Many members of the general public, and some academic scientists as well, maintain that at least some UFO sightings result from the activities of extraterrestrial visitors. Recent polls show that approximately 57% of the public believes that UFOs are 'something real' as opposed to 'just people's imagination'. The figure rises to 70% belief for those who are less than 30 years old (Gallup, 1978), and have thus lived their entire lives in the age of television. UFO belief is not found predominantly only among the uneducated. A 1979 poll of its readers by Industrial Research and Development magazine shows that 61% believe that UFOs 'probably or definitely exist', a figure that rises to over 80% for those applied scientists and engineers under age 26. 'Outer space' is the most widely held explanation of their origin.

It is obviously true that even if the reality of UFOs were somehow to be fully established, it would not prove the reality of extraterrestrial visitors. UFOs could possibly be, for example, some poorly understood atmospheric phenomenon, or the result of some secret terrestrial technology, or even a life form or natural phenomenon which lies totally beyond the scope of present-day science. But in the public mind, the subject of UFOs is inextricably linked with the idea of extraterrestrial intelligent life, and since ETI is the subject matter of this book, I shall henceforth adopt the popular usage of terms, and examine UFO reports in the context of the evidence they purport to contain concerning extraterrestrial visitors.

Although even the strongest proponents of the reality of UFOs concede that the vast majority of reported UFO incidents are the result of the misidentification of conventional objects, they maintain that after the elimination of this noise there remain a reasonable number of observations of UFOs by reliable observers which can only be explained in terms of alien visitors.

However, when we take into account the obvious fallibility of human eyewitness testimony, it is not surprising that there would be a very small residue of supposedly 'unexplainable' cases. The police do not achieve a 100% solution rate of armed robberies or hit-and-run accidents, yet no reasonable person would suggest that this proves that alien beings have been robbing banks and running down pedestrians. Allan Hendry of the Center for UFO Studies published a comparison of the more than 90% IFOs – Identifiable Flying Objects – reported to the Center over a one-year period, with the less than 10% of the reports which apparently defied identification (Hendry, 1979). His results were truly remarkable. He found that there was no significant statistical difference between the control group of misidentified objects – the IFOs – and the supposedly real UFOs. The two categories were virtually indistinguishable in terms of the duration of the occurrence, the time of day, the age and sex breakdowns of the witnesses, their occupational background, and their previous UFO involvement and interest. (It may be quite significant that both groups, the IFO reporters and the UFO reporters, reveal a degree of prior UFO interest which seems orders of magnitudes higher than that for the general population, although we have no data at this time to make an exact comparison.)

Hendry's data also reveal that, of the more than 200 reports he received that can be unquestionably confirmed as sightings of nocturnal advertising aircraft, more than 90% of the witnesses did not describe what was perceptually available to them, but instead imagined that they saw a rotating, disk-shaped form. Ten percent of these witnesses imagined that they could also see a dome on top of the nonexistent 'saucer' (Hendry, 1978). Hendry's quantitative evaluation of UFO data has confirmed what many of us have suspected all along: that there is no significant statistical difference between IFO and supposed UFO reports, that raw reports as received by investigative groups are frequently filled with gross errors of observation, and that in a small percentage of cases these inaccuracies are so overwhelming as to totally preclude any rational explanation when the report is taken at face value.

When subjected to a careful and detailed investigation, many of those cases which are frequently cited by UFO proponents as being the most convincing turn out to be readily susceptible to conventional explanation. Perhaps the best-known UFO incident on record concerns the supposed 'UFO abduction' of a New Hampshire couple, Barney and Betty Hill. While driving through a sparsely populated region of the White Mountains on the night of 19–20 September 1961, they allegedly encountered a mysterious aerial object which supposedly interfered with their conscious recollections, and somehow caused them to 'lose' 2 hours from their journey. When the Hills later underwent psychiatric treatment by Dr Benjamin Simon, a noted Boston psychiatrist, they both told under hypnosis similar stories of being supposedly 'abducted' by humanoid alien creatures, and being given a physical examination aboard a mysterious craft. The Hills' account has been the subject of a best-selling book and a made-for-TV movie, was serialized in a major national magazine, and has been cited dozens of times by supposedly scientific UFO investigators as evidence of the reality of alien visitors.

What the UFO proponents nearly always fail to mention, however, is that the Hills' psychiatrist, Dr Simon, has unambiguously stated on several occasions that, in his professional judgement, the 'UFO abduction' story represents a fantasy, and not a real event (Klass, 1974). Mrs Hill had originally experienced a series of dreams whose content was essentially identical with the story she later told under hypnosis. She described these dreams to anyone who wanted to listen – there were many who did – and her husband, of course, heard her narration many times. Hypnosis is not, we must remember, a road to absolute truth. A person suffering from a delusion might recite that delusion under hypnosis with great sincerity. I should point out that Mrs Hill's description of the supposed configuration of the UFO and a star near the moon that evening sounds remarkably similar to the Moon–Saturn–Jupiter pattern that actually existed. Had a genuine UFO been present, she would have seen three starlike objects in the vicinity of the Moon: Jupiter, Saturn *and* the UFO. But she saw only two.

Much has been made of an 'alien star map' that Mrs Hill reportedly saw aboard the UFO, and sketched afterwards. Some persons, including a few with impressive academic credentials, have endorsed an analysis purporting to show that the pattern of dots drawn by Mrs Hill can be uniquely identified as a carefully selected subset of the nearby stars as seen from a certain perspective outside our solar system, chosen on the basis of parameters that would make them prime candidates for supporting habitable planets (Dickinson, 1974). However, I have elsewhere published an analysis of the inconsistencies and ad hoc procedures required to achieve this supposed match (Sheaffer, 1981). Astronomers Steven Soter and Carl Sagan have noted that, in the absence of a grid of lines drawn in to suggest the relationship, no two patterns could be more dissimilar than the Hill sketch and the main-sequence star pattern (Soter & Sagan, 1975). Perhaps worst of all, there have been three mutually exclusive purported 'identifications' of stars supposedly represented in the Hill sketch published so far, each one having one or more features said to argue in its favor.

In recent years, Mrs Hill claims to have discovered a UFO 'landing spot' in New Hampshire, where she goes as often as three times a week to watch the UFOs land. She claims that the aliens sometimes shoot down beams at her, including one that 'blistered the paint' on her car (Hill, 1978). On other occasions, she alleges aliens have peeped into bedroom windows and have got out of their parked saucer to do callisthenics before getting back in again (*Sentinel*, 1978). UFO investigator John Oswald, who is certainly not a debunker, accompanied

Mrs Hill to the 'landing spot'. He reported upon his return that Mrs Hill was 'unable to distinguish between a landed UFO and a streetlight'. He nonetheless still believes that she was abducted by aliens in 1961 (*Skeptical Inquirer*, 1978).

Another famous UFO incident concerns the supposed near-landing of a UFO on a farm near Delphos, Kansas, in 1971. This case was selected as the most impressive out of a collection of more than 1000 incidents by a panel of UFO researchers holding PhD's, headed by the late astronomer J. Allen Hynek. Not only was there the testimony of apparently credible witnesses, but a whitish powdery ring, whose origin and strange properties seemed to defy analysis, was to be seen on the ground where the UFO allegedly had hovered. Skeptical UFO investigator Philip J. Klass travelled to the site, and interviewed the witnesses. He documented the farmer making several wild-sounding claims which were later shown to be false (Klass, 1974). The supposedly mysterious ring, which was for several years touted as convincing UFO evidence and to which was attributed totally bizarre physical and physiological properties, was recently disclosed with utterly no fanfare to have been finally identified by a French biologist as a growth of one of the fungus-like organisms Actinomycetaceae, of the genus *Nocardia*. Information scientist Jacques Vallee suggests, however, that high-energy stimulation from a UFO may have triggered the growth of the *Nocardia* (Vallee, 1975). The credibility of this case is further eroded by the claims of the principal witness, the farmer's son, that the UFO was responsible for virgin births among animals on the farm, and that he later sighted 'the Wolf Girl', who reportedly ran across the moonlit field on all fours 'faster than anything human can run' (Sheaffer, 1981).

I could go on to cite several dozen more UFO cases, similar to these, which have risen to prominence. But as this is to be a short chapter, the above will have to suffice. I think that all reasonable persons will have to agree that the credibility of the cases described above is essentially nil. The fact that cases like these are touted in pro-UFO writings and endorsed by the few pro-UFO scientists tells us a great deal about the lack of critical thinking which unfortunately is the norm in UFO circles. It also tells us a great deal about the credibility of the numerous lesser-known UFO cases, which must presumably be even less.

For centuries, one of the cornerstones of scientific methodology has been the principle commonly known as Occam's Razor: 'Essences are not to be multiplied beyond necessity', or, in the vernacular, extraordinary hypotheses are not to be invoked until all ordinary ones have been conclusively eliminated. For example, Occam's Razor would have prevented any truly scientific UFO researcher from writing that a fungus growth in the vicinity of an alleged UFO landing was likely caused by energy emitted from the UFO, unless it could be convincingly demonstrated that such a fungus growth is virtually impossible in ordinary circumstances.

When the evidence offered in favor of the reality of UFOs is critically examined according to the dictates of Occam's Razor, it is clear that there is little or nothing remaining which merits scientific scrutiny, because there is simply no UFO evidence yet presented that is not readily attributable to prosaic causes. The great bulk of what has been offered as evidence for the reality of UFOs consists of unsubstantiated statements by one or more witnesses. We have already noted the observational errors which are rampant in UFO reports. Recent experiments by psychologist Elizabeth Loftus underscore the inherent unreliability of unsubstantiated human observation and recollection (Loftus, 1979). Furthermore, Hendry's data contain the somewhat surprising result that observations reported by multiple witnesses are in fact slightly less reliable than those by a single witness (Hendry, 1979). Perhaps the psychologists can give us some insight as to why that is the case.

Many purported UFO photographs have been published, but in every instance either the object is insufficiently distinct, or the circumstances surrounding the incident are insufficiently credible, rendering the photograph unconvincing as evidence for anything. Occasionally, tangible effects of a UFO's presence are reported – the so-called 'Close Encounters of the Second Kind'. But, again, nothing has occurred that convincingly rules out prosaic causes. Electromagnetic interference effects have been attributed to the supposed proximity of UFOs. But such effects are sporadic and curiously localized, and always disappear without a trace before they can be verified. In instances where permanent magnetic imprints would be expected if the report was accurate, none have been found (Condon, 1969). The permanent and tangible effects attributed to UFOs consist almost exclusively of things such as broken tree branches, fungus-like rings or growths on the ground, and the death of cattle under allegedly unusual circumstances.

When the only evidence one can accumulate is of such dubious character, Occam's Razor leaves one no choice but to reject any and all remarkable explanations for the supposed UFO phenomenon and to attribute it to prosaic causes. The 'null hypothesis' has not been excluded.

In recent decades many short-lived or rare phenomena have successfully been photographed or detected unambiguously by scientific instruments. Yet UFOs seem to have an utterly infallible ability to avoid unambiguous detection. Alert photographers have captured such brief, unexpected events as commercial airliners falling from the sky, and brilliant daylight meteors. Yet where is the corresponding unambiguous photograph of the often-reported daylight disk UFO? This lack of evidence is all the more puzzling when it is remembered that there is no shortage of UFO reports from major populated areas. We have supposedly authentic close-encounter cases from such places as the suburban areas of Chicago and Pittsburgh, the crowded Santa Ana Freeway near Los Angeles at noontime, and even near the New Jersey entrance of the George Washington Bridge into Manhattan.

Several years ago I coined the term 'jealous phenomena' to denote those alleged but never-proven phenomena which have the ability to flawlessly play peek-a-boo with reality. They are defined as those phenomena which *always* manage to slip away before the evidence becomes too convincing. Joining UFOs are such other jealous phenomena as ESP, Bigfoot sightings, psychic spoon bending and the Loch Ness monster.

The astonishing elusiveness of the supposed UFO phenomenon, and the undeniably dream-like character of many of the reports, have caused even many advocates of the reality of UFOs to abandon the hypothesis of extraterrestrial visitors in favor of even more bizarre hypotheses. J. Allen Hynek has suggested that UFOs may originate in some as yet unknown parallel plane of existence, or may represent non-physical visitations from elsewhere. In a 1976 interview he explained that 'perhaps an advanced civilization understands the interaction between mind and matter – in the manner of the Geller phenomena, for instance – much better than we do. Perhaps somebody out there is able to project a thought form and materialize it down here, à la "Star Trek". There are other planes of existence – the astral plane, the etheric plane, and so forth' (Hynek, 1976). Jacques Vallee has noted the remarkable similarity between reports of UFO occupants and reports of fairy sightings of an earlier age. His most recent of many UFO hypotheses is that UFO sightings are caused by some secret human organization for as yet unknown purposes of deception, using secret high-technology psychic devices (Vallee, 1979). He does not attempt to reconcile this latest interpretation with his earlier hypothesis that sightings of UFO occupants represent the same phenomenon as earlier sightings of fairies, and, hence, these alleged secret psychotronic devices must have been in existence at least since the Middle Ages (Vallee, 1969). Popular UFO writer John Keel attributes the UFO phenomenon to beings he calls 'ultraterrestrials', who, he says, are 'our next-door neighbors, part of another space–time continuum where life, matter, and energy are radically different from ours' (Keel, 1970).

Thus, if you, the reader, have always found it next to impossible to give any credence to the extraterrestrial hypothesis for UFOs, owing to the extreme difficulty of wide-ranging interstellar travel and the implausible character of the reports themselves, it may come as a surprise to learn that there is a significant number of UFO proponents who have come to share your opinion. But, unfortunately for this avant-garde wing of saucerdom, their ill-defined ideas about 'parallel universes' and 'psychic interaction' cannot in any way hope to face up to Occam's Razor.

Another factor which argues against the extraterrestrial or other anomalistic explanation of UFOs is the fact that the percentage of supposed unknowns does not vary with the overall level of sightings. The percentage of supposed unidentifieds is *not* an invariant, as some UFO researchers have stated, but the percentage varies according to factors other than the overall level of UFO sightings. As an illustration, let us consider some data from the records of the US Air Force's Project Bluebook. The year 1963 was an extremely quiet one for UFOs; 1966 was a year of great excitement, with nearly three times as many total reports. Yet the percentage of supposed unidentified in both years was approximately 3%. The number of sightings in 1968 was more than double the number of the following year, yet the percentage of supposed unexplaineds was exactly the same, seven-tenths of 1%. Sightings increased by 50% from 1956 to 1957, yet the percentage of supposed unknowns stayed virtually the same. Sightings jumped a dramatic 788% from 1951 to 1952, and while the percentage of supposedly genuine UFOs increased from 12.6% to 19.3%, that is a much smaller change than would be expected from the nine-fold increase in sightings. Indeed, one suspects that the Air Force's investigators may have simply been overwhelmed by the volume of reports in 1952 and were unable to investigate them adequately, because if we rely instead on a revised analysis of the Bluebook files (Hynek, 1977), we find that, despite the 788% increase in reports from 1951 to 1952, the percentage of supposed unidentifieds was essentially the same.

This is a most puzzling factor for the following reasons. Suppose that the unidentified UFO reports represent sightings of alien spacecraft. Then, when the number of genuine UFO sightings triples in a given year, it is presumably because the ones that are here have become three times as active. One would expect the percentage of supposed unidentified to go up dramatically, as the signal-to-noise ratio improves, but they do not. Why should a three-fold increase in extraterrestrial activity cause people to also report Venus and weather balloons as UFOs at three times the previous rate? How did the Venus spotters determine so accurately the rate of increase of genuine UFO activity that they were able to keep the IFO/UFO ratio essentially unchanged? Or, looking at it from the other side, suppose that the rate of sightings of genuine alien spacecraft does not change significantly from year to year, but that the remarkable 'UFO flaps' as we know them are manifestations of a mild form of mass hysteria. If so, one would expect that the percentage of genuine UFOs would *decrease* dramatically during a 'flap' as the noise increases but the signal remains the same. Yet this does not happen. Do the supposedly genuine alien spacecraft therefore deliberately increase their activities in proportion to the hysteria to keep the IFO/UFO ratio essentially unchanged?

I suggest that the most straightforward explanation of the above dilemma is that the signal-to-noise ratio in UFO reports is exactly zero, and that the apparently unexplainable residue is due to the essentially random nature of gross misperception and misreporting. This would explain why, no matter how many or how few UFOs are reported, the IFO/UFO ratio does not vary accordingly. If the supposedly genuine UFOs represent nothing more than that minority of cases in which the 'random noise' factor in the human perceptual and conceptual apparatus becomes so large as to totally overwhelm the original stimulus for the observation, then this would explain why the percentage of supposedly real UFOs is essentially independent of rises and falls in the overall number of reports.

Of course, none of the above factors in any way proves that extraterrestrial visitations are not now taking place. Such a proof is impossible. One could no more prove the non-occurrence of such a thing than one could prove that there are no dinosaurs left alive on earth, or that witches do not fly on broomsticks. But what has been established, in my judgement quite convincingly, is that there is no evidence in favor of extraterrestrial visitations that comes anywhere close to establishing that claim as a viable scientific hypothesis. To paraphrase Samuel Johnson, we have no other reason for doubting the existence of extraterrestrial visitors than we have for doubting the existence of men with three heads: very simply, we do not know that there are any such things. As has been said many times, extraordinary claims require extraordinary proof, and the burden of proof always rests on the person who asserts the existence of an anomaly. Until such proof is supplied, the null hypothesis remains unrefuted, and the existence of extraterrestrial visitors is not in any way suggested or established.

References

CONDON, E. U. (ed.), (1969), *Scientific Study of Unidentified Flying Objects*. pp. 100; 282 (case 12); 380 (case 39); 749. New York: Bantam.

DICKINSON, T. (1974). *Astronomy*, 2 (no. 12), 4.

GALLUP, G. (1978). Opinion poll, May, 1978.

HENDRY, A. (1978). *International UFO Reporter*, 3 (no. 6), 6.

HENDRY, A. (1979). *The UFO Handbook*, chapters 2–8, 20. New York: Doubleday.

HILL, B. (1978). *UFO Report*, January.

HYNEK, J. (1976). *Fate*, June.

HYNEK, J. (1977). *The Hynek UFO Report*. pp. 253–267. New York: Dell.

KEEL, J. (1970). *UFOs Operation Trojan Horse*, chapter 15. New York: Putnam.

KLASS, P. J. (1974). *UFOs Explained*, pp. 253; 312–332. New York: Random House.

LOFTUS, E. F. (1979). *American Scientist*, 67, 312.

Sentinel (1978). News story of 27 June. Centralia, Illinois.

SHEAFFER, R. (1981). *The UFO Verdict*, chapters 5, 18. Buffalo, New York: Prometheus Books.

Skeptical Inquirer (1978). Vol. 3 (No. 1), 14.

SOTER, S. & SAGAN, C. (1975). *Astronomy*, **3** (No. 7), 39.
VALLEE, J. (1969). *Passport to Magonia*. Chicago: Regnery.
VALLEE, J. (1975). *The Invisible College*, chapter 1. New York: Dutton.
VALLEE, J. (1979). *Messengers of Deception*. Berkeley, California: And/Or Press.

4

The Likelihood of Interstellar Colonization, and the Absence of Its Evidence

SEBASTIAN VON HOERNER

According to several estimates, up to 0.5% of all stars could have a planet similar to our Earth, but on the average about four billion years older than Earth, because our Sun is not an old star and star formation was most productive in the early times. Regarding the origin and evolution of life, our own case is at present the only instance of life we know of. Are we permitted to generalize this single case? Can we do statistics with $n = 1$? The laws of *statistics* say that $n = 1$ yields an estimate for the average, but none for the mean error (which would need at least $n = 2$). This means that assuming us to be average has the highest probability of being right, but we do not have any indication of how wrong this may be. Leaving statistics and arguing by *analogy*, we may add that most things in nature do not scatter over too large a range, up to a few powers of ten, mostly. Thus, the best we can do is to assume that we are average, but to allow for a wide (but not infinite) error of this assumption. If we now generalize our own case, then life in our Galaxy would have started on about one billion planets several billion years ago. And, arguing by *extrapolation*, we should expect this life to have developed meanwhile extremely far beyond our own present state. However, nothing but *wild* guesses can be made regarding the direction, the range and possibly the termination of such far-out developments. What activities should we then expect that would be visible to us? And mainly: Why don't we see any? This is a great, tantalizing puzzle.

Looking at our own present activities leads us back to Frank Drake's old question 'Is There Intelligent Life on Earth?' This is one of the real values of SETI: it makes us look at Earth from a distance and in general terms. Our own large-scale public activities have mostly been self-destructive. At present, world-wide military expenditures are about 800 billion dollars per year, with almost no decline after the end of the 'cold war' escalation, which, by the way, was not ended by intelligent reason at all but by the unreasonable economic ruin of one of the two rivals. The destructive power piled up in nuclear bombs is now 3000 times larger than all explosives used in 5½ years of World War II (which killed about 45 million people). And, divided by our world population, the destructive nuclear power equals 2 tons of TNT per capita; which means 4000 pounds of dynamite for every man, woman and child on Earth, black or white or yellow. And if we generalize our case, we may expect that a goodly percentage of all technical civilizations have ended, undetected, by self-destruction. But hardly all of them. Technology may act as a filter, allowing into a future only those who have developed wisdom together with intelligence.

Provided we *do* have a future, let us guess some of our future activities. We are now using up our terrestrial resources at an alarmingly fast and still increasing rate, and within a few more generations we must develop, introduce and enforce rather elaborate and expensive recycling systems for most of the metals and some other minerals; but no recycling system can ever work 100%. We have just now begun to feel an energy crisis, the really dramatic part of which is still ahead of us. But if demands are increasing and supplies are running out, why on Earth should we stay?

The following chapters will show that large self-sustaining, growing and multiplying colonies throughout our solar system can be made, for mining and production, even with our present beginner's technology and with less money than the suicidal arms race. After some large original investments, the colonies will become quite vital and profitable when our resources on Earth run out. These colonies will grow with their own babies and grandchildren, and, after more and more generations out in space, their people will feel less and less of their originally strong ties to the home planet Earth, probably up to some Declarations of Independence.

Larger groups of thousands of volunteers may decide to take off in huge 'mobile homes' on interstellar trips lasting many generations, finally colonizing the planets of other stars. And after a certain settling time on such a planet, the same cycle may repeat, leading again to mobile homes on another interstellar trip to the next stars and their planets. In this way we would have started a wave of stepwise colonizations, finally covering the whole Galaxy from one end to the other, and every nice planet in it. The complete galactic colonization could well be finished within some 10 million years, even with our present limited knowledge of physics and technology.

And here we have our great puzzle. Such a wave of colonization could have been started by any one out of the billion early civilizations in our Galaxy. Our Earth should have been colonized long ago, and we ourselves should be the descendants of some early settlers, and not the homegrown humans that we certainly are.

Should we really assume such an urge for interstellar colonization; is this reasonable and justified? First, it is only necessary that at least one civilization felt the urge strong enough to get it done, out of a billion potential ones. Furthermore, quite in general, life shows a strong tendency to fill out every possible niche, from dry deserts to cold polar regions, from swamps to bare rocks, from caves to the tops of mountains. Life has started in the water, it has conquered first the land and soon the air; it begins right now to conquer nearby empty space, and so it may quite naturally proceed to conquer interstellar distances as well.

What enables or triggers the larger steps of this development? The decisive milestones of evolution are set by introducing and exploiting new ways of information handling. All life, self-reproductive life, began with the genetic code, which is a most ingenious way of storing and reproducing all the information needed for storing and reproducing that very information plus whatever is needed for growing the whole organism surrounding it, including information about maintenance and repair and behavior. The next large step, the development of higher life, was made possible by growing a nervous system with a brain as its main office of command, where incoming informations are evaluated, memories are stored and outgoing instructions are given. The third large step, our whole human culture, is based on the development of speech about a million years ago. It has been drastically enhanced by the introduction of script, and now it begins another revolution by using cybernetic means. And so this whole evolution should quite naturally proceed to a large-scale network of interstellar communication, involving all the many members of the Galactic Club. And again we have our puzzle: that all this should actually have been done long ago, that the whole Galaxy should be teeming with life, that 'empty' space should be bristling with messages and probes, some of it obvious in many ways, whereas we have not yet found any evidence of any extraterrestrials.

Many reasons against colonization have been mentioned for explaining the puzzle. For example: self-destruction of technologies, biological degeneration, stagnation by over-stabilization against crises, complete change of cultural interests, space technology never becoming cost-effective, a repetition time for travelling and settling of much more than a million years, colonization turning from an organized procedure into a random walk. All of these reasons could very well hold in some cases, maybe even in most cases, but hardly with no exception at all in a billion. What matters is the far-out tail of a highly populated distribution.

A possible conclusion, then, is that we are actually alone, by some reason not yet understood. Or could it just be that one in a billion of nice habitable planets has been overlooked or neglected by the colonizers? And regarding all the other evidence to be expected: maybe we do not look for the right thing, or we do not understand what we see? The great puzzle is still unsolved.

Maybe we are just too impatient, expecting results after only 30 years of some occasional SETI searches. The duration of search needed for success cannot be guessed if there is a Galactic Club, which may shorten the time by beaming strong signals at us, or lengthen it by neglecting us (disgusted by our TV and news). But I could estimate the minimum search time, for the first ones long ago who ever tried to communicate. We have two unknowns, the needed length L of time and the distance D to be covered. And we have two equations. First, nothing goes faster than light; thus, the time between sending a signal and getting an answer, even with *perfect* methods, is at least twice the distance divided by the speed of light. Second, the longer the average search duration, the less the distance to the nearest partner who tries simultaneously. The result is a time of $L = 5000$ years and a distance of $D = 2500$ light-years (where I assumed that the maximum fraction of all stars where communication is ever tried for L years is 1%, but its uncertainty enters the resulting L and D only with the power $\frac{1}{4}$.)

This holds for the first founders of a Galactic Club. Once a network has been established, its further growth can go much faster. Still, success for us may need a substantial dedicated search of some duration. And, as has already been said many times *we shall never know if we do not try*. Even no success after a long dedicated search would be an important answer, regarding our own role in this universe. Well, success or not, I seriously hope that SETI will help to widen our horizon and to open our mind. A Chinese colleague once said: Mind is like parachute, works best when open.

Bibliography

BALL, J. A. (1973). *Icarus*, **19**, 347.

BOVA, B. (1963). *Amaz. Fact. Sci. Fict.* **37**, 113.

BRACEWELL, R. N. (1974). *The Galactic Club*. San Francisco: W. H. Freeman.

CAMERON, A. G. W. (ed.), (1963), *Interstellar Communication*, New York: Benjamin.

DOLE, S. H. (1970). *Habitable Planets for Man*. New York: Elsevier.

DRAKE, F. & SOBELL, D. (1992). *Is Anyone out There?* New York: Delacorte Press.

DYSON, F. J. (1968). *Physics Today*, **21**, 41.

HART, M. H. (1975). *Quart. J. Royal Astronomical Soc.*, **16**, 128.

HEIDMANN, J. & KLEIN, M. J. (eds.) (1991). *Bioastronomy*. Berlin: Springer-Verlag.

HOERNER, S. VON (1975). *J. Brit. Interplanetary Soc.*, **28**, 691.

HOERNER, S. VON (1978). *Naturwissenschaften*, **65**, 553.

JONES, E. M. (1976). *Icarus*, **28**, 421.

O'NEILL, G. K. (1974). *Physics Today*, **27**, (No. 9), 32.

O'NEILL, G. K. (1977). *The High Frontier*. New York: William Morrow.

PAPAGIANNIS, M. D. (1977). International Conference on the Origin of Life, Japan. *Astron. Contrib. Boston Univ.*, Series II (No. 61).

SAGAN, C. (ed.) (1973). *Communication with Extraterrestrial Intelligence*. Cambridge, Mass.: MIT Press.

SAMPSON, A. (1977). *The Arms Bazaar*. New York: Viking Press.

SIVARD, R. L. (1993). *World Military and Social Expenditure*. Washington, DC: World Priorities.

5

Pre-emption of the Galaxy by the First Advanced Civilization

RONALD BRACEWELL

In one school of thought it is customary to begin discussions of galactic life by appeal to Drake's equation and then to proceed to a detailed examination of the numerical magnitude of one or more of the string of factors whose values have to be estimated. An example of this procedure is furnished by Michael H. Hart's analysis (Chapter 22), in which he concentrates on the probability that 600 or more nucleotides might line up in the right order; then he proposes that one of the factors may be very much less than 10^{-30}. Of course, 10^{-30} is already very small, and, if included as a factor in almost any expression having to do with the physical universe, will cut the product down to negligible size. In this application the conclusion is that the number of technological civilizations independently arising in a galaxy is very much less than 1. Well, this may be an excess of zeal, and many of those addicted to the use of Drake's equation would, in similar circumstances, have arranged for the product to emerge with an order of magnitude around unity because, after all, a calculation condemns itself if it seriously contradicts the possibility of the one technological civilization we know about, namely our own.

But is Drake's equation correct? It seems that it suffers from oversimplification – surely at least one plus sign ought to be there. This matter was taken up at greater length in connection with a meeting of the International Astronautical Federation (Bracewell, 1979a) and the conclusion is that the equation has to be generalized. At the very least, more than one route to life and certainly to intelligent life, has to be allowed for. As the cited paper states, 'One would not attempt to estimate N_f, the number of fish in the sea, by multiplying an estimate of R_f, the rate of formation of fish, by L_f, the average longevity of a fish. There are too many *kinds* of fish to make this simplification *useful*.' There are other troubles with the scientific foundations of Drake's equation.

My inclination is to admit that the plurality of technological civilizations is an open question and to follow up the different logical possibilities with a view to discerning courses of action that we might take. Thus, it is conceivable that the Galaxy teems with life – in that case radio listening makes sense and will soon provide confirmation. Happily that kind of action is already under way. But life is conceivably much more rare. If life is too sparse, radio listening might be fruitless, but there might be something else we could do. As is well known, a proposal was made in 1960 (Bracewell, 1960) that in such a case, referred to as Case II, we should be alert to the possible appearance of a messenger probe. At that time it seemed sensible, on a basis of cost-effectiveness, that such an alien messenger probe ought not to bear the penalty of having to make a planetary landing on unfamiliar terrain but should simply go into orbit in the habitable zone of the Sun and explore the radio spectrum for indications of technological activity. But now it seems to me that, under Case II, some principle of equipartition of effort ought to apply. A move toward such a principle already underlies the Cyclops philosophy, where, however, it is understood that the alien civilization maintaining the powerful radio beacon over millennia is making a contribution that far outweighs the terrestrial contribution, which is merely to construct and man a listening post for perhaps a decade; even so, the Cyclops design would strain national resources.

The equipartition principle, then, means that the participants make efforts that, in terms of their capacities, are comparable. Applying this principle to contact by messenger probe, we ask ourselves: What effort is it that Earth could make that exceeds the role of mere alertness, as originally suggested, and that substantially enhances the effectiveness of the probe? One answer is that we could relieve the probe of the need to orbit the Sun by maintaining a solar system space watch for a probe that may simply be flying by. Such a probe would approach at a great speed and we would have to detect it in the depths of space and act quickly. A velocity of several tens of kilometers per second could be expected – probably something in excess of the escape velocity from the solar system. Even so, as we know from the behavior of comets, such as Halley's comet, that have come from great distances, the time of passage through the inner solar system is many days. This should be sufficient to extract the message content from a cooperative probe. Whether there would be a bonus to be gained from the greater effort on our part that would be entailed in a rendezvous with the probe, and whether such a rendezvous would be depended upon by the designers of the probe, seems doubtful to me because the lodging of a message is paramount and can be done by radio without need of physical contact, fascinating though that could be.

Under this more equitable sharing of effort it would not be necessary to equip the probe with the very substantial retrorocket needed for halting here, and as a result more probes and more reliable ones could be launched under whatever constraints were operative. The above discussion is to illustrate my theme of identifying separate courses of action appropriate to various logical possibilities as to density of technological life. Of course, in both cases so far mentioned technological life exists: in Case I it is dense and in Case II it is less dense.

There is another good illustration of this action-oriented theme which applies even if life is sparse, in which case one may at least seek planets. There are various ways of searching (Bracewell, 1979b), all of which require to be developed simultaneously. A scheme using an apodized space telescope worked out by B. M. Oliver is very promising and needs to be followed up. Another scheme proposes an interferometer (Bracewell, 1978; Bracewell & MacPhie, 1979) spinning in space and makes use of infrared radiation from the hypothetical planet because such radiation is 10^5 times stronger relative to that from the parent star than is the case with visible light. Both the space-based schemes involve advanced technology that does not yet exist, but there are no apparent fundamental limitations, and the prospects of technical success are favorable. In this illustration the action indicated is development of appropriate space technology. We are reminded that, where technology (as distinct from science) is involved, certain decisions are better postponed pending the outcome of preliminary studies and necessary experimental development.

Among the logical possibilities is one that has not been much favored but which in my opinion requires equal attention, again with a view to discerning appropriate action. I take the orthodox scientific position of not making up my mind until there is evidence. To some people such fence-sitting is uncomfortable – they like to make up their minds and to be known as decisive and nonstodgy people. In 1964 I wrote a well-anthologized chapter entitled 'Are We Alone?' and it wasn't long before Walter Sullivan took a liking to this phrase and brought out a book entitled *We Are Not Alone*. Well, I don't know how he knows that we are not alone. We very well may be alone, but this has not been a favored hypothesis, even though it leads to a most interesting discussion. It is Case IV. (It should be mentioned that Case III refers to a situation where technological life in the Galaxy is so sparse that under secular equilibrium conditions the longevity of a civilization is less than the round trip travel time of electromagnetic radiation to the nearest neighbor. The conclusion that longevity has an inverse relation to the spacing of neighbors (Bracewell, 1960) under equilibrium conditions is now generally accepted.)

Here is the reasoning connected with Case IV. Intelligent life on earth evolved through competition under definite conditions of time and place that might, as far as we know, have arisen at other times or in other phases on earth, but did not. Our arboreal ancestors in Africa developed the precise binocular color vision that we benefit from today, together with the manual dexterity and coordination of hand and eye that are desirable in leaping from branch to branch. On descending to the forest floor, our ancestors developed other human skills, tool making and speech, possibly under the demand for weaponry and advance planning for group action fostered by big game hunting. Be these details as they may.

Why didn't a tree kangaroo descend to the Australian savannah, adopt an upright posture, move to a mixed diet that included occasional birds' eggs and lizards, and go on to group hunting of giant macropods? Honing his intelligence on internecine warfare with related kangaroo species, an intelligent marsupial might have emerged accompanied by no relatives closer than the peaceful grazing kangaroos of today, which perhaps might have become the basis of a meat and leather industry. Similarly, why didn't the three-toed sloth descend onto the South American pampas or the racoon emerge onto the prairies to hunt the buffalo. There are any number of creatures with abilities comparable with those of our arboreal ancestors, many living in geographic conditions not obviously different from what Africa offered. Why haven't they progressed further on the upward path to civilization?

Regardless of the detailed reasons that may have impeded such progress in the past, we can now relate what in fact did happen once man passed the threshold of mobility, because he migrated over the whole Earth and by his presence now preempts the possibility of future evolution in directions that would compete with his supremacy. Signs of intelligence would bring immediate retribution, and presumably the absence of any gradations between ourselves and the chimpanzee is due to harsh suppression that occurred at some time in the past when the struggle to determine which intelligent strains would survive was still unsettled.

It is not customary in biology to speak as if Nature had a plan, but it leads to convenient phraseology. If there was a plan to populate the Earth with intelligent creatures, it did not depend on trying out different schemes at different times and in different places. On the contrary, Nature's plan was to do a good job in just one place. It took a lot of time. But when a critical threshold was passed, that successful progenitor moved out to cover the whole world. Very possibly some further physical evolution took place along the way, but all in the same line of descent – no bears or racoons were called in once the primate line paid off. So, when the age of world exploration began, men found all the continents except Antarctica and many remote islands to be already populated by intelligent beings – their own relatives.

The fact that the Earth is populated with intelligent creatures is not because the many habitable areas of Earth were able to foster the evolution of intelligence: it is because one area was the scene of the events; and the rest of the Earth is now a habitat of intelligence, courtesy of Africa. Even Antarctica is now inhabited by man.

In time to come, perhaps not too far off, there will be colonies on the Moon and in space, comparable with the many colonies now maintained in Antarctica. We shall then have to admit that cislunar space is populated, and then in due course Mars and perhaps other planets and satellites will be. Until recently it was considered quite possible that life might have originated independently on Mars, and probably this possibility is not entirely abandoned even today, but it is now clear that intelligent life will not evolve on Mars. Even if the weather were to improve, the possibility of Mars being left intact for the necessary eons has been preempted by the appearance of man. Mars may indeed come to be populated, but not by Martians.

When this comes to pass we could confirm that Nature's plan to populate the solar system had indeed been to develop a good mobile model in one place and have it migrate to other habitats.

As we can already foresee the first unmanned exploration to nearby star systems, it is natural to turn now to speculation about the Galaxy as a whole. Perhaps you think Nature's plan is to have man fill the Galaxy following the pattern of the continents and the foreseeable pattern for the planets. But I do not believe you can so conclude before asking why it was that the plan worked on Earth. There is a very simple mathematical consideration that sheds light on this question, an idea that was first published in the San Francisco *Examiner*, 29 June 1975, where it could have had no effect on scientists except those on the Berkeley–Stanford axis. The idea is simply that humans could walk from Africa to California, long and hard though the trek may have been, in *less time* than it would take for intelligence to evolve independently in California or elsewhere, and therefore that the key question is: *How long would it take to migrate through the Galaxy as compared with evolution time?* The conclusion was: 'At moderate speeds of space travel the human race can reach the centre of the Galaxy in much less time than the 3.5 billion years that it took the earth to produce man.' (Bracewell, 1975)

This is a fascinating conclusion. It does not say that we *will* do it but it says that we *can* do it. It changes our self-estimate from what Cyclops implies, namely that we are inferior, to the view that we may have a noble destiny. Let's follow up this view.

We are profoundly cosmic creatures. The majority of the atoms in our bodies are hydrogen atoms that were created in the Big Bang 15 billion years ago. Most of the mass of our bodies is in oxygen atoms that were created in the generation of stars, now gone, that preceded the formation of our Sun. Human protoplasm is continually being created from the lifeless mineral constituents of our planet. We are a subset of the material of the physical universe and now we are confronted by the spectacle that a negligible subset of the universe is, through astronomy, acquiring an awareness, albeit imperfect, of the total universe. How can the part be aware of the whole? But not only that, we are beginning to understand the universe and to bring parts of it under control. At the same time we are converting the inanimate physical matter into protoplasm.

Now we are on the threshold of distributing our protoplasm, and the intelligence and awareness that go with it, back into the cosmic space which is where we came from, from the stars. We are in a real sense part and parcel of the Galaxy, not mere observers, onlookers. Perhaps our role will be to influence the physical evolution of the Galaxy, bringing to galactic evolution the phenomenon of consciousness which already influences man's own progress and, though as yet geographically very confined, is nevertheless a part of physics. If we *are* alone, we may be on the threshold of a magnificent destiny.

References

BRACEWELL, R. N. (1960). Communications from superior galactic communities. *Nature*, **186**, 670.

BRACEWELL, R. N. (1975). Other voices. *San Francisco Sunday Examiner and Chronicle*, B3 (29 June issue).

BRACEWELL, R. N. (1978). Detecting nonsolar planets by spinning infrared interferometer. *Nature*, **274**, 780.

BRACEWELL, R. N. (1979a). An extended Drake's equation, the longevity–separation relation, equilibrium, inhomogeneities and chain formation. *Acta Astronautica*, **6**, 67–9.

BRACEWELL, R. N. (1979b). Life in outer space. *Proc. Roy. Soc. New South Wales*, **112**, 139.

BRACEWELL, R. N. & MACPHIE, R. H. (1979). Searching for nonsolar planets. *Icarus*, **38**, 136.

6

Stellar Evolution: Motivation for Mass Interstellar Migrations*

BEN ZUCKERMAN

Introduction

One of the most controversial aspects of the problem of life in the universe is the value of N, the number of technological civilizations that exist in an average spiral galaxy such as the Milky Way. N has been debated at various meetings (e.g. Papagiannis, 1980) and extreme values between 10^{10} and 10^{-24} have been suggested. One of the strongest arguments in favor of small N is the so-called 'Fermi paradox': If N is a large number, then why are extraterrestrials not physically present in our solar system (see, e.g., Hart & Zuckerman, 1982, hereafter HZ)? Various arguments have been advanced to explain this paradox and yet allow a large value of N. For example, Drake (see Papagiannis, 1980, p. 27) has contended that it is not cost-effective to travel between the stars using rocket ships and, therefore, even if N is a large number, the extraterrestrials will choose to stay home. The question of cost-effectiveness is a debatable one, in any event (e.g. Singer in HZ, p. 46).

The purpose of this chapter is to point out that, if N is large, then, for a wide class of reasonable scenarios, extensive rocket travel between the stars seems not only likely but inevitable, quite independent of considerations of cost-effectiveness, speed of colonization waves, etc. The basic reason is that large N implies that L, the lifetime of an average technological civilization, must be very long, at least millions of years. It seems inconceivable that a civilization that has lasted for this long will, then, somehow cease to exist (owing to a catastrophic event, for example). Therefore, such venerable civilizations should last at least as long as the main-sequence lifetime of their respective 'Suns'.

* Reprinted from the *Quarterly Journal of The Royal Astronomical Society* 26, 56–9, (1985), with permission of Blackwell Scientific Publications, Ltd.

At this point the extraterrestrials have two choices: they can attempt to adjust to a relatively rapidly changing environment as their Sun becomes a red giant and then a white dwarf, or they can all emigrate to other stars. The luminosity of the white dwarf will, typically, be at least 1000 times smaller than the luminosity of its main-sequence progenitor. Since the white dwarf can, therefore, support only a tiny fraction of the population that could exist near the main-sequence or red giant star, the former possibility seems very unlikely indeed.

We show below that if $N \sim 10 - 100$, then at least one civilization has probably been forced to migrate. If $N \sim 10^5$, as some astronomers have estimated, then the entire Milky Way should, by now, have been extensively (completely?) colonized by civilizations that originated around stars that have already left the main sequence.

Stellar Evolution and Mass Migration

We want to calculate the number of technological civilizations that have originated on planets that orbit around main-sequence stars which have evolved in a time that is less than the age of our Galaxy, which we assume to be 10^{10} yr. We assume a standard Salpeter initial mass function with dQ/dt equal to the rate of formation of stars in the Galaxy with masses between M and $M + dM$. Then:

$$\frac{dQ}{dt} = 1.3 \frac{dM}{M^{2.3}}$$

where, following Iben (1984), we have normalized to a galactic birthrate of one star per year with a mass greater than the mass of our Sun (M_{Sun}). Integration of this birthrate, assumed constant over the past 10^{10} yr, yields the total number of stars with masses between 0.1 and 1 M_{Sun}:

$$Q_{total} = \int_0^{10^{10}} dt \int_{0.1}^1 1.3 \frac{dM}{M^{2.3}} = 2 \times 10^{11} \text{ stars}$$

More massive stars add less than 10% of this total.

Stars with nearly the mass of the Sun have main-sequence lifetimes that vary like M^{-3}. We assume that stars with main-sequence lifetimes that are shorter than 5×10^9 yr ($M > 1.26 \, M_{Sun}$) do not last sufficiently long to permit a technological civilization to evolve on planets orbiting around them. Stars with mass less than M_{Sun} have main-sequence lifetimes $>10^{10}$ yr and have not yet had time to evolve off the main sequence. Therefore, the interesting range of stellar masses lies between 1 and 1.26 M_{Sun}. The total number of stars in the Milky Way born in this mass range is:

$$Q_i = 10^{10} \int_1^{1.26} \frac{1.3 \; dM}{M^{2.3}} = 2.6 \times 10^9 \; \text{stars}$$

The number of these stars still on the main sequence is:

$$Q_a = \int_1^{1.26} 1.3 \, \frac{dM}{M^{2.3}} \int_0^{10^{10}/M^3} dt = 1.9 \times 10^9 \; \text{stars}$$

So the number of stars that have 'died' is:

$$Q_i - Q_a = 7 \times 10^8$$

The stars that have died were formed from interstellar clouds that had fewer 'metals' (i.e. all elements heavier than helium) than did our Sun. One might argue that just after our Galaxy formed there were insufficient metals to form planets around solar mass stars. Theories of the origin of planetary systems are sufficiently rudimentary to prevent our really knowing the minimum metal abundance that is required for the formation of planets. In any event, considerations such as these seem unlikely to change our estimates by more than a factor of 2 (see, e.g., Trimble in HZ, p. 135), which is but a very minor uncertainty compared with various other assumptions.

Now we need to estimate how many civilizations (N^*) might have existed on planets orbiting around these 7×10^8 'dead' stars. Calculations of N are usually placed into the context of the 'Drake equation' (see, e.g., Goldsmith & Owen 1980; HZ). This equation may be written simply as:

$$N = \alpha L$$

where α lumps together all factors other than the lifetime of the technological civilization. Estimates for α range from 300 to 10^{-11} yr^{-1} (Goldsmith & Owen 1980) or even to much smaller values (Hart in HZ, p. 154). We argued in the introductory section that all technological civilizations that manage to exist for a 'long time' will then, in fact, last for billions of years. With 'best estimates' for α by Goldsmith & Owen and by Sagan (table 18.1 in Goldsmith & Owen, 1980) the implied value of N is ~10^9, which seems very large even by optimistic standards. Therefore, to match a more conventional value for N (~10^5), we may assume α = 4×10^{-5}, where L is assumed to equal 2.5×10^9 yr (i.e. steady birth of technological civilizations over the past 5×10^9 yr with no destruction).

If we assume that technological civilizations form with equal probability around all stars with masses between 0.1 and 1.26 M_{Sun}, then the minimum number of civilizations that have existed on planets orbiting around stars that have left the main sequence is approximately

$$N^*_{min} = \frac{7 \times 10^8}{10^{11}} N \sim 10^{-2} N$$

since, with our above assumptions, about one-half of the 2×10^{11} stars in this mass range were born within the past 5×10^9 yr and cannot yet have given birth to technological civilizations. Most astronomers would agree, however, that stars with $1 < M < 1.26\,M_{Sun}$ are more likely to harbor indigenous intelligent civilizations than the more numerous M-type dwarfs having $M \sim 0.2\,M_{Sun}$. Perhaps the more massive stars are favored by a factor of 10 over the stars of lower mass (see, e.g., Hart in HZ, p. 154). In this case, a maximum estimate for N^* is:

$$N^*_{max} \sim 10^{-1}\,N$$

So we see that even if N is as small as 10–100, at least one venerable civilization will have had to face the termination of the main-sequence evolution of its home star. If $N = 10^5$, then more than 10^3 civilizations will have met this fate.

What might such a civilization do? As we have suggested above, massive migration seems the most likely possibility, although it is perhaps a bit presumptuous for us to speculate on how a 10^9-year-old civilization might react to a crisis! It is perhaps equally difficult to imagine just how such a civilization might be organized. Here we assume that the advanced civilization is utilizing its material and energy resources to a reasonably high degree and has constructed many millions of space colonies, each one of which holds a substantial number of creatures (see, e.g., Singer in HZ, p. 46).

The termination of the life of a star largely vitiates considerations of cost-effectiveness and it is hard to imagine that these colonies will not propel themselves to nearby stars. Because a typical star such as our Sun passes within 100 light-years of $\sim 10^6$ stars during the 250 Myr that it takes to orbit once around the centre of the Milky Way, and because advanced civilizations could certainly sense their impending crisis, it seems probable that they will send out colonies over an extended period of time and thus populate at least 10^6 other star systems. If $N^* \sim 10^4$, the Galaxy may well have been completely filled by these mass migrations.

Conclusions

One might argue that the above paradigm is filled with assumptions about the actions of very advanced civilizations. However, if one believes that N is large and accepts the conventional picture of stellar and galactic evolution, our conclusions seem almost inevitable. Alternatives seem much less plausible. For example, one might argue that zero population growth will be achieved in the near future (let us hope that it will be) and our descendants will never build the very many space colonies envisaged by proponents of Dyson spheres and Kardashev Type

II civilizations. Even if this is the case, i.e. our descendants do not take apart the entire solar system to build colonies to support a maximal population, it is, nonetheless, hard to believe that, in the next few billion years, they will not utilize ('improve') at least a few otherwise useless items such as the asteroids. The largest asteroid, by itself, could easily supply the necessary raw materials for more than 10^6 large space colonies.

Other explanations that purport to explain why extraterrestrials are not now here on Earth, such as the Zoo Hypothesis (Ball, 1973), seem equally implausible in the face of mass migrations. In a crisis such as the death of one's beloved home star, would the affected society worry about preserving 'wilderness areas'?

We believe that the above discussion illustrates, yet once again, the insurmountable problems inherent in the construction of a logically consistent model of galactic evolution whereby N is large and yet the only (or even primary) means of communication between different star systems is electromagnetic radiation.

Acknowledgements

I thank my colleagues Drs M. Jura and M. Morris for suggesting that I should publish these ideas and for commenting on the manuscript.

References

BALL, J. A. (1973). *Icarus*, **19**, 347.
GOLDSMITH, D. & OWEN, T. (1980). *The Search for Life in the Universe*. Menlo Park, California: Benjamin/Cummings.
HART, M. H. & ZUCKERMAN, B. (1982). *Extraterrestrials, Where Are They?* New York: Pergamon Press.
IBEN, I. (1984). In *IAU Symp. No. 105*. Dordrecht: Reidel.
PAPAGIANNIS, M. D. (1980). *Strategies for the Search for Life in the Universe*. Dordrecht: Reidel.

7

Interstellar Propulsion Systems

FREEMAN DYSON

There is no lack of propulsion systems available to any creatures which possess some technical competence and a desire to travel around in the galaxy. The following is an incomplete list of propulsion systems which have been suggested and studied by members of our own species.

Group A: Systems which are certainly feasible but are limited to mission velocities of the order of $10^{-2}c$.

1. Nuclear-electric. Uses a fission reactor as energy source, and ion-beam or magneto-hydrodynamic plasma jet for propulsion. One can imagine a 'minimal starship' using nuclear-electric propulsion, with 10^{-6} g acceleration, a mass of 5 kg/kW (electric) for reactor and radiator, and a mission duration of 10^4 yr for voyages of the order of 10 pc.
2. Old-fashioned Orion nuclear pulse propulsion, using full-sized fission or fusion bombs. This is also limited to velocities of the order of $10^{-2}c$ but can have acceleration of the order of 1g, giving it much better performance in local maneuvers (see Dyson, 1968; Martin & Bond, 1979).

Group B: Systems which are probably feasible but require very demanding new technology. These systems should be capable of mission velocities of the order of 0.5c, and mission durations of a few decades for distances of a few parsecs.

3. Laser-driven sails (see Norem, 1969).
4. Microwave-driven sails (see Forward, 1985).
5. Pellet-stream propulsion (see Singer, 1980).
6. Direct electromagnetic launch (see Clarke, 1950).

The laser-driven sail system requires a laser with output power of 10^{13} W, emitted coherently over an aperture of the order of 30 km diameter.

Forward's 'Starwisp' proposal for a microwave-driven sail requires a transmitter power of only 10^{10} W. The total mass of Starwisp is 20 g. If it were required to carry human passengers, the size and transmitter power of Starwisp would become comparable with the size and power of the laser-driven sail.

The pellet-stream system uses a transmission-line magnetic accelerator or rail-

45

gun to accelerate solid pellets to moderately relativistic velocities. The ship has the job of intercepting and absorbing momentum from the pellets. If the pellets can be accelerated at 10^4g, the length of the gun will be of the order of 10^8 km.

Direct electromagnetic launch uses a similar rail-gun to accelerate a space vehicle in one piece. This would be a preferred alternative if the passengers in the vehicle could survive high accelerations.

Group C: Systems which I shall not discuss, because I consider them to be inferior either in feasibility or in performance to Groups A and B.

7. Project Daedalus (JBIS, 1978).
8. Bussard Ram-Jet (Bussard, 1960).
9. Celestial Billiards (Ford, 1959).
10. Matter–Antimatter Annihilation Rocket (Steigman, 1974).

For further references to all these systems, see the bibliographies by Mallove & Forward (1972) and Mallove *et al.* (1977).

The most efficient and economical systems are those of Group B, which have the energy source located in a massive fixed installation decoupled from the vehicle. The vehicle, not needing to carry its own energy source, can be made small and light. These are the only systems which offer travel at relativistic speeds with reasonable efficiency in mass and energy.

Unfortunately, the Group B systems are better at accelerating than at decelerating. Norem has discussed a method of deceleration for the laser-driven sail, using the deflection of electrostatic charge by the interstellar magnetic field to turn the vehicle velocity around through 180 degrees, so that the vehicle can decelerate using the same laser beam that was used to accelerate it. A similarly complicated maneuver has been suggested by Singer for deceleration in the pellet-stream system. These deceleration maneuvers greatly increase the size and complexity of the launch systems and are useless for voyages extending more than a few parsecs from the launcher. So the main conclusion which follows from this examination of interstellar propulsion systems is the following.

We Have The Accelerator! How About The Brakes?

The Group B systems would provide versatile and efficient rapid transit around the Galaxy, provided that there existed a correspondingly efficient and versatile braking system. By a 'braking system' I mean a self-contained device by means of which a fast-moving vehicle could slow down and stop, transferring its momentum to the interstellar plasma. I know of only one candidate for a braking system, namely:

Group D: Braking systems.

11. Alfven Propulsion Engine (see Drell *et al.*, 1965). The Alfven Propulsion Engine was discovered accidentally when it turned out that the ECHO I satellite (a large balloon orbiting the earth at 1600 km altitude) experienced a drag force exceeding the expected aerodynamic drag by a factor of about 50. Drell and coworkers explained the anomalous drag as the effect of electromagnetic coupling between the vehicle and the plasma in the Earth's magnetosphere. The momentum of the vehicle was efficiently transferred to the plasma by Alfven waves traveling along the magnetic field-lines.

The theory of Drell *et al.* gives a simple formula for the Alfven drag:

$$D_a = \rho\, V V_a\, A \tag{7.1}$$

where ρ is the mass density of the plasma, V the vehicle velocity, A the frontal area of the vehicle, and V_a the Alfven velocity,

$$V_a = \sqrt{B^2/4\pi\rho} \tag{7.2}$$

with B the magnetic field component transverse to the flight path. The theory was derived from Maxwell's equations and verified by the ECHO satellite only for the case of sub-Alfven vehicle velocity

$$V << V_a \tag{7.3}$$

In the earth's magnetosphere V_a is of the order of 100 km/s and the theory is certainly applicable. Unfortunately, the Alfven velocity in interstellar plasma is only of the order of 10 km/s, and the interstellar vehicles will be in the super-Alfven regime

$$V >> V_a \tag{7.4}$$

It is an urgent theoretical problem to investigate whether the drag formula (7.1) still applies in the super-Alfven regime.

Another way of writing the Alfven drag formula (7.1) is

$$D_a = \sqrt{D_h D_m} \tag{7.5}$$

where D_h is the ordinary hydrodynamic drag

$$D_h = \rho\, V^2\, A \tag{7.6}$$

and D_m is the magnetic drag

$$D_m = (B^2/4\,\pi)\, A \tag{7.7}$$

In the earth's magnetosphere the true drag D_a is larger than the hydrodynamic drag D_h. I am conjecturing that in the super-Alfven regime the true drag is still D_a, although it is then much less than D_h.

If (7.1) holds, then the deceleration of a vehicle of mass m is given by

$$m\dot{V} = -\rho V V_a A \tag{7.8}$$

The vehicle comes to rest exponentially fast, with a characteristic stopping time

$$\tau = \frac{m}{\rho V_a A} \qquad (7.9)$$

A typical interstellar medium has ρV_a of the order of 10^{-18} g cm^{-2} s^{-1}. To give a stopping time of the order of a few years, we need a vehicle mass per unit area

$$\frac{m}{A} \leq 10^{-10} \text{ g/cm}^2 \qquad (7.10)$$

A mass per unit area of 10^{-10} g/cm^2 would be absurd for a continuous surface, but it is perhaps not absurd for a braking system. The braking system need not cover its area continuously, but could consist of thin long wires spread far apart. For example, it could carry current in a network of 1 μm diameter wires spaced 10 m apart. Detailed analysis is needed to find out whether such a network could be made electrically conducting and mechanically strong enough to decelerate itself by coupling to the interstellar plasma.

If it turns out that interstellar braking systems are feasible, then we have a new way to look for evidence of extraterrestrial intelligence. Look for skid-marks on the road! A vehicle braking from high velocity will leave behind it a long straight trail of hot plasma which should be a source of persistent broadband radio emission. Radioastronomers interested in CETI should be on the look-out for straight tracks of glowing plasma in the sky. They should also be careful to make sure their signal-to-noise ratio is high enough, so that their tracks do not prove as illusory as Lowell's Martian Canals.

References

BUSSARD, R. W. (1960). *Astronautica Acta*, **6**, 179–94.

CLARKE, A. C. (1950). *J. Brit. Interplanetary Soc.*, **9**, 261–7.

DRELL, S. D., FOLEY, H. M. & RUDERMAN, M. A. (1965). *Physical Review Letters*, **14**, 171–5.

DYSON, F. (1968). *Physics Today* (October issue), 41–5.

FORD, K. W. (1959). Los Alamos T-Division Report.

FORWARD, R. L. (1985). *J. Spacecraft Rockets*, **22**, 345–50.

JBIS (1978). J. *Brit. Interplanetary Soc., Interstellar Studies Supplement*.

MALLOVE, E. F. & FORWARD, R. L. (1972). *Bibliography of Interstellar Travel and Communication – 1972*. Research Report 460, Hughes Research Labs, Malibu, California 90265.

MALLOVE, E. F., FORWARD, R. L. & PAPROTNY, Z. (1977). *Bibliography of Interstellar Travel and Communication, April 1977 Update*. Research Report 512, Hughes Research Labs, Malibu, California 90265.

MARTIN, A. R. & BOND, A. (1979). *J. Brit. Interplanetary Soc.*, **32**, 283–310.
NOREM, P. C. (June 1969). American Astronautical Soc., Paper 69–388.
SINGER, C. (1980). *J. Brit. Interplanetary Soc.* **33**, 107–15.
STEIGMAN, G. (1974). Yale University Preprint.

8
Interstellar Travel: A Review

IAN A. CRAWFORD

Introduction

There are both scientific and social reasons for wanting to go to the stars. On the scientific side, astronomy and planetary science (and very likely the biological sciences also) would benefit tremendously. Just consider the advantages of taking thermometers, magnetometers, mass spectrometers, gravimeters, seismometers, microscopes, and all the other paraphernalia of experimental science, to objects that today can only be observed telescopically across light-years of empty space. On the human side, it would seem that the total number of people who ultimately receive a chance of life, and the survival time of our species itself, would increase enormously if colonization of even a small part of the Galaxy were to prove possible. As pointed out by Shepherd (1952), 'humanity dispersed over many worlds would appear to be more secure than humanity crowded on one single planet'. At the very least, the resulting cultural diversity would provide an exciting alternative to Fukuyama's (1989) dire predictions for the 'end of history' (a point discussed in more detail by Crawford, 1993a).

In this chapter we review some of the propulsion methods that might make it possible to travel interstellar distances on a timescale of decades (i.e. velocities $\geq$ $0.1c$). The concepts discussed are necessarily selective, and the reader who wishes to dig deeper is referred to the extensive bibliography of interstellar travel and communication compiled by Mallove et al. (1980) and updated by Paprotny et al. (1984, 1986, 1987). Readers interested in the social consequences of interstellar travel are referred to the interesting book on the subject edited by Finney & Jones (1985).

The Pioneer and Voyager Interstellar Spacecraft

There are already four spacecraft on hyperbolic trajectories that will take them out of the solar system and into interstellar space. These are Pioneers 10 and 11 (launched in 1972 and 1973) and Voyagers 1 and 2 (launched in 1977), the interstellar trajectories of which have been described in detail by Cesarone *et al.* (1984). These four spacecraft have asymptotic heliocentric velocities (i.e. velocities relative to the sun once they have effectively escaped from its gravitational influence) in the range $10 - 17$ km s^{-1}. However, as 10 km s^{-1} corresponds to only 10^{-5}pc yr^{-1}, it is clear that much higher velocities will be required if interstellar travel is to occur on timescales relevant to human society. We should point out, however, that even these very low velocities may suffice for some proposed schemes of interstellar colonization (to which we will return briefly on p. 61), and are certainly sufficient for a program of directed panspermia such as that discussed by Crick (1981).

The Daedalus Project

There has, to date, been only one detailed engineering study of a starship design. This was the Daedalus Project, carried out between 1973 and 1978 under the auspices of the British Interplanetary Society. The aim was to design an automatic vehicle capable of travelling to Barnard's star (the next closest to the Sun after the α Centauri system, and for which, at that time, there seemed good circumstantial evidence for a planetary system) with a flight time of about 50 years. This would require a cruising speed of 12% that of light, which was to be achieved with currently available technology or a reasonable extrapolation thereof. If the same mission profile were adopted for a flight to the α Centauri system, the flight time would be 36 years. The payload was to have a mass of 450 t, and the only propulsion system capable of satisfying these criteria was found to be one based on nuclear fusion. Daedalus was to make an undecelerated fly-by of the target star system, it being quite impractical to *launch* a vehicle capable of *decelerating* a 450 t payload from 12% of the speed of light with the same propulsion system.

The vehicle that evolved from this study consisted of two stages, both of which were to utilize the fusion of deuterium and ^{3}He for propulsion. Pellets of these isotopes are injected into a reaction chamber, where they are heated and compressed to the point of thermonuclear ignition by relativistic electron beams. The resulting plasma is constrained by a magnetic field in such a way that it can escape in only one direction and therefore acts as a rocket exhaust. Full details of this propulsion system, and a review of its historical origins, are given by Martin & Bond (1978) and Bond & Martin (1978a); see also Martin & Bond (1979). Table 8.1 lists some characteristics of the two stages (Bond & Martin, 1978b; Strong & Bond, 1978).

Following the acceleration phase, the coast velocity of the second stage and payload would be 12.2% of the speed of light. There seems little doubt that

Table 8.1. *Daedalus vehicle specifications*

Parameter	First stage	Second stage
Length	140 m	110 m
Diameter (including propellant tanks)	190 m	80 m
Stage mass at cut-off	1690 t	980 t
Propellant mass	46 000 t	4 000 t
Propellant	D, ^{3}He	D, ^{3}He
Thrust	7.5×10^6 N	6.3×10^5 N
Burn duration	2.05 years	1.76 years

pulsed fusion propulsion systems, which may follow naturally from present inertial fusion research, would be capable of achieving these very high velocities. Of course, a very large quantity of nuclear fuel, and, hence, a very large vehicle, is required, and this has overawed some skeptics of the project. In order to retain a sense of proportion, it is helpful to bear in mind that we have no difficulty in building ships (supertankers, for example) which are far larger than Daedalus, and which could comfortably carry much more than 50 000 tonnes of fuel. The major problems involved in realizing Daedalus do not concern the size of the vehicle *per se*, but the fact that it must be constructed in space (something common to all starship concepts) and that the fuel consists of rare nuclear isotopes. ^{3}He, in particular, is very scarce, and the Daedalus study concluded that it would be necessary to 'mine' it from the helium-rich atmosphere of Jupiter (Parkinson, 1978). A recent review of the Daedalus project, including a discussion of points which have arisen since the publication of the original work, has been given by Bond & Martin (1986).

Before leaving the Daedalus concept, we shall consider the effect of interstellar matter on the vehicle. This was examined in detail by Martin (1978). At the Daedalus velocity an interstellar grain with a mass of 10^{-16} kg has a kinetic energy of about 0.07 J and, upon collision, is capable of producing very high temperatures at the point of impact (Benedikt, 1961). This will lead to evaporation of material from the vehicle and some form of shielding will be required. Ideal shield materials should have a low density, so as to minimize the shield mass, but large latent and specific heat capacities, so as to minimize the loss of material due to collisional heating. Martin concluded that beryllium, boron and graphite were the best available candidates. For a beryllium shield and a dust space density of 1.6×10^{-24} kg m^{-3} (as is probably appropriate for the solar neighborhood) Martin's analysis implies a total mass loss of 1.6 kg m^{-2} over the whole flight. This may be considered as an upper limit because, as pointed out by Powell (1975), much of the vaporized material will recondense onto the shield before it can escape into space.

Antimatter rockets

There are other possible nuclear propulsion systems in addition to the Daedalus concept. Rather than discuss these here (the interested reader is referred to the bibliographies referenced in the Introduction), we shall instead consider the potential performance of the most efficient possible rocket. The most efficient rocket fuel would consist of a mixture of matter and antimatter, as their mutual annihilation leads to a 100% conversion of mass into energy, although not all of this energy is actually released in a useful form. When a proton, p, and antiproton, $\bar{p}$, are annihilated, a shower of subatomic particles is produced. Initially, this consists of a number of charged and neutral pions, i.e.

$$p + \bar{p} \rightarrow n\pi^+ + n\pi^- + m\pi^0$$

such that, *on average*, $2n + m \sim 5$ (see, e.g., Vandermeulen, 1972). The neutral pions decay in about 8.3×10^{-17} s into two gamma rays. The charged pions decay (after $\approx 26 \times 10^{-9}$ s) into muons and neutrinos. The muons will then themselves decay into electrons, positrons and neutrinos after the much longer interval of 2.2 μs. Finally, if still physically associated, the electrons and positrons are annihilated to produce gamma rays.

A review of possible engine designs which could utilize these matter–antimatter annihilation products has been given by Morgan (1988). Basically, there are two ways in which the energy released by these reactions could be used to drive a rocket:

1. The charged annihilation products could be directed by a magnetic field, thereby producing thrust. A field strength of about 50 T is required if collimation is to be achieved within a space of about ½ m (Morgan 1982).
2. The annihilation products could be used to heat a much larger quantity of working fluid, which would then be expelled to produce thrust. This turns out to be a more efficient use of the antimatter, and will be considered in more detail below.

About 40% of the initial rest mass energy of the proton and antiproton is accounted for by the kinetic energy of the charged pions (see, e.g. Vulpetti 1986), and it is this energy which could, in principle, be transferred to a rocket exhaust. Of course, it will be necessary to contain the relativistic pions within a relatively small volume so that they transfer their kinetic energy to the working fluid before they decay. A number of possible confinement geometries are possible. For example, the pions could be constrained magnetically in a volume containing the working fluid, requiring a field strength of about 10 T for a reaction chamber of 1 m diameter (Morgan, 1982). Alternatively, they could interact directly with a liquid or solid propellant (Vulpetti, 1986), or be used to heat a solid surface past which the working fluid flows on its way to the rocket nozzle (Morgan, 1988, and references therein).

Regardless of exactly how the annihilation energy is passed to the working fluid, we can derive an expression for the quantity of antimatter required to achieve a particular mission. This derivation has been given previously by Shepherd (1952) and Dipprey (1975). For an overall efficiency of 40% in converting rest mass into exhaust kinetic energy (i.e. assuming that essentially all of the *kinetic* energy of the charged pions can be utilized), we find that the mass of antimatter required is given by

$$M_a \approx M_{veh} \times \left(\frac{\Delta v}{c}\right)^2 \tag{8.1}$$

where M_a is the mass of antimatter, M_{veh} is the *empty* mass of the vehicle (i.e. the mass of the payload, engine, structure, etc., but excluding the fuel mass), Δv is the velocity gained by the vehicle during its flight and c is the speed of light. This quantity of antimatter must be used to heat a much larger quantity of inert reaction fluid, M_{rf}, the optimum quantity of which turns out to be given by

$$M_{rf} \approx 3.9 \times M_{veh} \tag{8.2}$$

It follows from eqs. (8.1) and (8.2) that if we wished to achieve a velocity increment of $0.1c$, we would require 10 kg of antimatter, and 3.9 t of reaction fluid, for each tonne of empty vehicle mass.

As the empty vehicle mass includes the engine and the rocket structure, in addition to the scientific payload, these must be taken into account when considering the quantity of antimatter required for a particular mission. It is difficult to estimate a plausible mass for an interstellar antimatter engine; however, since it is likely to include equipment for the generation of powerful magnetic fields and for the cryogenic confinement of antimatter, as well as at least some radiation shielding, it seems clear that several tonnes will be involved. For the sake of argument, we shall assume an engine mass of 20 t (which may be wildly optimistic) and, following Cassenti (1982), we shall further assume that the remaining structural mass amounts to 5% of the mass of reaction fluid.

Suppose we wish to accelerate a nominal 1 t payload to 10% of the speed of light. Equation (8.2) (rearranged so as to include the engine and structural masses) shows that we shall require 102 t of reaction fluid and eq. (8.1) tells us that we require 260 kg of antimatter. However, most of this is required to accelerate the relatively large engine mass, so there is actually little to be gained in having a payload mass small compared with that of the engine. On the other hand, payloads more massive than the engine will dominate the economics, e.g. the Daedalus mission (M_{pl} = 450 t, Δv = 0.12c) would require 2280 t of reaction fluid and 8.4 t of antimatter.

If we wished to bring the 1 t payload to rest from its cruise velocity of 0.1c, we would require a first stage capable of accelerating to 0.1c all that is necessary for the deceleration phase. Assuming that the same engine is used, we find that 620 t of reaction fluid and 1.6 t of antimatter are required for the acceleration phase. Note that the total quantity of reaction fluid needed to accelerate a 1 t payload to 0.1c and bring it to rest again (722 t) is comparable to the quantity of liquid propellant required by the space shuttle to achieve Earth orbit!

We see that, even for a comparatively modest interstellar mission, the mass of antimatter required will be of the order of several hundred kilograms. Needless to say, there will be many difficulties involved in producing, cooling and storing these macroscopic quantities of antimatter. These problems, together with some proposed solutions, have been discussed by, among others, Forward (1982), Mitchell (1988), Stwalley (1988) and Michaelis & Bingham (1988). For the sake of argument, we shall take the optimistic view that if sufficient antiprotons can be produced, technical solutions will be found to the problems of producing solid antihydrogen in macroscopic quantities.

At the present time, antiprotons are produced in physics laboratories by impacting a beam of high-energy protons onto a metal target (see von Egidy, 1987, for a review). The kinetic energy of the protons in the beam goes into the formation of a large number of nuclear particles, some of which are antiprotons. The antiproton collector (ACOL) at CERN is currently able to collect about 10^{12} antiprotons per day from the target. If ACOL were to run continuously, this rate of collection would yield about 6×10^{-7} mg of antimatter per year. Several studies (e.g. Larson, 1988) have been carried out in order to establish the technical requirements of an antiproton factory capable of producing 1 mg yr^{-1} (i.e. 2×10^{13} antiprotons per second), and it appears that foreseeable technology may be capable of achieving this rate of antiproton production.

Thus, the quantity of antimatter required for use as an interstellar rocket fuel far exceeds our present production capabilities. Nevertheless, in order to gain some insight into what will be required in order to produce kilograms of antimatter per year, consider an antihydrogen factory based on a 100 TW proton beam (Forward, 1982; 1TW = 10^{12} W). There would be enough raw energy in such a beam to create 1 g of antiprotons per second. However, antiproton production is very inefficient, partly because of the inevitable production of less massive particles, and partly because existing techniques can only collect a small fraction of the antiprotons created at the target. The figures given by Goldman (1988) imply an overall efficiency in converting primary proton beam energy into antiproton rest mass of about 10^{-7}. As we are considering a dedicated antiproton factory, let us assume that the efficiency could be increased to about 10^{-4}. Such a facility would then be capable of producing about 3 kg of antiprotons per year. Moreover, it may be that, as our knowledge of particle physics increases, ways will be found to suppress the formation of less massive particles at the expense of antiprotons, thereby increasing the efficiency still further. In this respect, it is interesting that Baldin *et al.* (1988) have reported that the ratio of antiprotons to pions produced by collision of a carbon nucleus with a copper target is 60 times higher than that produced by proton collisions with the same energy per nucleon. A further efficiency increase of this order would enable our hypothetical accelerator to produce 180 kg of antiprotons per year.

The energy required for antimatter production is placed in perspective when we realize that the present-day electrical generating capacity of the whole world is of the order of only 1 TW! Clearly, it is not, and may never be, practical to generate the required energy at the surface of the Earth. From an astronomical perspective, however, 100 TW is actually a very small amount of power (about $2.6 \times 10^{-13} L_{Sun}$), and at the Earth's distance from the Sun would be intercepted by a collector only 270 km on a side. Thus, we see that the Sun is the obvious source of energy for this (and other) proposed methods of interstellar travel.

Having discussed the performance of the most efficient possible rocket, we now turn to proposed methods of interstellar transport which rely on other principles entirely.

Light Sails

Photons carry a momentum given by $h\nu/c$, where h is Planck's constant, ν is the frequency of the light and c is the speed of light. It follows that a beam of light could, in principle, be used to accelerate a space vehicle, and it has long been recognized that spacecraft deploying large reflective surfaces ('sails') would be able to utilize the momentum carried by sunlight as a means of getting around the inner solar system. In order for the method to be adapted to interstellar flight, however, it will be necessary to arrange for a powerful unidirectional beam of light (as produced by a laser, for example) in order to avoid the inverse-square dependence of the intensity of sunlight on distance. This method of interstellar propulsion was first proposed by Forward (1962), and subsequently considered by, among others, Marx (1966), Redding (1967), Moeckel (1972) and Forward (1984).

If we are interested in using photon pressure to accelerate a vehicle to a significant fraction of the speed of light in a few years, a very intense beam of light will be required. For a perfectly reflecting surface, the acceleration resulting from the change of momentum of the incident photons is given by

$$a = \frac{2P}{Mc} \qquad (8.3)$$

where P is the power in the beam and M is the mass of the vehicle. For example, if we wanted to send a 450 t payload to Barnard's star in 50 years (the nominal Daedalus mission) we would require a constant acceleration of 0.045 m s^{-2}, corresponding to a radiated power of 3×10^{12} W.

In practice, many other considerations enter into the design of an interstellar light sail. Foremost among these are (1) the requirement that the incident power be distributed over a sufficiently large area to prevent structural damage, and (2) the fact that even a perfectly collimated beam will diverge because of the optical phenomenon of diffraction. Both of these considerations result in the need for a very large reflective area, i.e. a light sail, even though one is not required by a simple application of eq. (8.3).

By considering the equilibrium between the incoming power P and the energy radiated by the sail (from both surfaces), it can be shown that the vehicle acceleration and the temperature of the sail are related by

$$a = \frac{2(1 + \eta)\sigma \, \varepsilon T^4}{c} \frac{A}{\alpha \, M} \qquad (8.4)$$

where σ is Stephan's constant; ε is the sail emissivity; η, α, are the sail reflectance and absorptance; T is the temperature of the sail; A is the sail area; and M is the total vehicle mass (payload plus sail). (In deriving eq. (8.4) we have substituted $(1 + \eta)$ for the factor 2 in eq. (8.3) in order to allow for the fact that not all the incident radiation is reflected.)

The sail thickness must be chosen to maximize the acceleration and minimize the amount of light that passes through the sail. Forward (1984) discusses a number of possible sail materials and finds aluminum to be the most appropriate, for which the optimum thickness is 16 nm, assuming a wavelength of 650 nm for the incident light. For a thin sail the maximum operating temperature is far below the melting temperature because of the danger of agglomeration, and it appears that the maximum safe temperature for an aluminum sail would be about 600 K (Forward, 1984). Thus in what follows we shall adopt values appropriate to a 16-nm-thick aluminium sail: $\varepsilon = 0.06$, $\eta = 0.82$, $\alpha = 0.135$, $\rho = 2700$ kg m^{-3}, $T = 600$ K. We shall also assume that the vehicle has a structural mass (i.e. mass of sail supports, stays, etc.) which equals the mass of the reflective surface itself.

Table 8.2. *Vehicle parameters for the three light sail missions discussed in the text*

Payload mass (t)	Sail diameter (km)	Total vehicle mass (t)	Power (TW)	Transmitter diameter (km)
1	6.9	4.2	0.24	440
450	147	1 920	110	21
3500	410	14 900	860	7.4

The power required to achieve a given acceleration can be obtained from eq. (8.3), taking into account the size (and, hence, mass) of sail required (and again substituting $(1 + \eta)$ for the factor 2). For example, if we wanted to send a 450 t payload on a flyby mission to α Centauri in 36 years we would require a continuous acceleration of about 0.06 m s^{-2}. Equations 3 and 4 show that we would require a sail diameter of 32 km (with a mass of 34 t, excluding structural support) and a total transmitted power of 5.1×10^{12} W. The final velocity at Centauri would be 23% of the speed of light. The same mission flown with 1 t and 3500 t payloads would require sail diameters of 1.5 and 88 km, and transmitted powers of 1.2×10^{10} and 4.0×10^{13} W, respectively.

As diffraction will lead to divergence of the beam, very large transmitting apertures will be required. The diameter of the central peak in the diffraction pattern formed at a distance, s, from a circular aperture of diameter D is given by

$$d = 2.44 \, s\lambda/D \qquad (8.5)$$

where λ is the wavelength of light employed. For $\lambda = 650$ nm, it follows that aperture diameters of 42 000 km, 2000 km and 720 km would be required to produce beam diameters equal to the sail diameters of the 1 t, 450 t and 3500 t payload missions at the distance of the nearest star.

One way to reduce the size of the transmitting aperture would be to employ higher accelerations, and cease power transmission when the vehicle has traveled only a fraction of its total flight path. For example, the same 36 year flight to the nearest star could be achieved with an initial acceleration of 0.35 m s^{-2} lasting for 3.3 yr, by which time the velocity would be 12% of the speed of light (the same as envisaged for the Daedalus vehicle) and the spacecraft would be 0.06 parsecs from the solar system. The sail diameters, total vehicle masses, transmitted powers and transmitting aperture diameters required in this case are summarized in table 8.2.

The transmitting apertures required are much larger than could plausibly be constructed for a single laser, and Forward (1984) has suggested the construction of a Fresnel zone lens of the appropriate size in the outer solar system. This would collect the individual beams from a number of separate lasers situated closer to the Sun. It is also clear that very large power levels are required, typically many TW for all but the most modest missions. As in the case of antimatter production, it appears that this energy must ultimately come from the Sun.

Before leaving interstellar light sails, we must briefly consider the effect of interstellar material on the sail. Considering that 1.6 kg m^{-2} was found to be eroded from the Daedalus dust shield over its 1.8 pc flight to Barnard's star, and that a 16-nm-thick aluminum sail would have a surface density of only 4.3×10^{-5} kg m^{-2}, the prospects for interstellar light sails would at first sight appear hopeless. However, Forward (1986) has argued that as the sail is so thin (much thinner than the ~0.1 μm diameter of interstellar grains), the grains will pass straight through the sail, depositing little of their kinetic energy within it. If we follow this assumption, and further assume a local interstellar grain density of 10^{-13} grains cm^{-3} and a grain radius of 0.1 μm, the fraction of the sail surface converted into holes is ~10^{-4} for each parsec travelled. Since we have shown above that, in order to minimize transmitter size, the acceleration phase would be best restricted to within less than a parsec of the Sun, interstellar grains would appear not to be a significant problem for light sails operating in the solar neighborhood.

The Laser Powered Interstellar Rocket

Rather than using the momentum of photons to push a light sail toward the stars, it may be possible to use the energy transmitted by a solar system based laser to heat a quantity of inert reaction fluid carried by the spacecraft. A relativistic treatment of such a vehicle has been given by Jackson & Whitmire (1978). Here we give a simple non-relativistic treatment which follows that given above for antimatter rockets. The two cases are similar, since, in the antimatter case, the mass of *energy* carried by the vehicle is negligible, while in this case the vehicle carries *zero* energy, as this is all transmitted from outside. The mass of reaction fluid required by a laser powered spacecraft is therefore the same as that given by eq. (8.2). As for the energy required, if E is the total energy beamed to the vehicle, and this is converted into exhaust kinetic energy with an efficiency ε, we obtain

$$E = 0.772 \frac{M_{veh} \Delta v^2}{\varepsilon} \qquad (8.6)$$

Except for the very lightest payloads, eq. (8.6) indicates that a laser powered rocket is a more efficient use of laser energy than a light sail. For example, the total energy required to accelerate the Daedalus 450 t payload to 12% of the speed of light by this method (assuming $\varepsilon = 0.5$) is 1.2×10^{21} J (where, as for the antimatter case, I have assumed a 20 t engine and a structural mass equal to 5% of the reaction fluid mass, yielding $M_{veh} = 584$ t). This compares with a total energy of 1.1×10^{22} J released over the 3.3 yr acceleration phase for the appropriate entry in table 8.2.

It is also interesting to compare this propulsion system with the antimatter rocket. Recall that the production of antimatter is very inefficient (it was speculated above that this efficiency might possibly be made to approach 10^{-4} in a dedicated antimatter factory), so it is actually orders of magnitude more efficient to beam the solar energy to a laser powered rocket than to use it to create antimatter as an intermediate step. However, the concept still suffers from the problem of maintaining a highly collimated beam over interstellar distances.

The Interstellar Pellet Stream

Singer (1980) has suggested that a stream of electromagnetically launched pellets might be used to transfer momentum to an interstellar vehicle. Physically this concept is similar to the light sail, except that material pellets (with a mass of a few grams and velocity $\sim 0.2c$) are used to transfer momentum instead of photons. As with the light sail, the major difficulty appears to be collimation of the pellet stream, although Singer has shown that it may be possible to correct the course of each pellet while in flight, something that is not possible with photons.

The pellet stream concept does not escape from the problems of ambitious engineering, as the electromagnetic launcher will probably have to be tens of thousands of kilometers in length in order to achieve the desired pellet velocities. However, if the engineering difficulties of the launcher can be overcome, and pellet collimation maintained, Singer was able to show that this concept offers considerable advantages over both fusion rockets and laser pushed light sails. As an example of the potential performance of the pellet stream technique, Singer considered its application to the nominal Daedalus mission: in order to accelerate a 450 t payload to $0.12c$ in 3.8 yr, a maximum launcher power of 15 TW is required. This is about half of that generated by the Daedalus first stage, and almost an order of magnitude less than required by a laser pushed light sail operating with the same acceleration distance.

People to the Stars?

It will always be more difficult to send people to the stars than machines, and some proposed solutions to this greater problem are discussed in this and the following section. There appear to be three ways in which human beings might travel to the stars.

World ships   This concept envisions very large vehicles designed to carry a breeding population of human beings over interstellar distances. These vehicles would employ some form of nuclear propulsion to achieve velocities of $\leq 10^{-2}c$. The travel times would, of course, be very long (≥ 1000 yr), the idea being that the remote descendants of the original pioneers would one day arrive at a destination. A review of the world ship concept, with references to its historical development, has been given by Martin (1984). As an example of the scale of vehicle considered, Bond & Martin (1984) have considered the construction of a cylindrical vehicle 20 km in diameter and 114 km in length. The cylinder is spun about its long axis so as to provide artificial gravity, and is designed such that its interior surface is as Earth-like as possible. Interesting moral and social problems arise when one considers the many generations that would be destined to live and die within a world ship before it reached any other stellar system, and these have been discussed by Holmes (1984) and Regis (1985).

Suspended animation   We have seen that a number of foreseeable propulsion technologies (e.g. fusion, laser, and antimatter drives) may be capable of achieving velocities of 10–20% of the speed of light. These velocities correspond to travel times of several decades for journeys to the nearer stars, and such a significant fraction of a lifespan spent in transit is unlikely to be acceptable to a human crew. However, if some way can be found of drastically slowing down human metabolic processes, involving loss of consciousness as well as delayed aging, there is no reason why travel times of decades would be unacceptable from a human point of view. An interesting review of biological and medical work which may have a bearing on suspended animation for spaceflight has been given by Hands (1985). There appears to be no fundamental reason why the human metabolism should not be greatly slowed down for several decades (or perhaps even stopped and then restarted?), although the necessary techniques do not exist at present.

Relativistic time dilation   If an interstellar vehicle were to travel very close to the speed of light, relativistic time dilation would enable its crew to travel great distances in little time as measured in their frame of reference. An interval of ship time, Δt_s, is related to an interval of Earth time, Δt_0, through the well-known relation:

$$\Delta t_s = \sqrt{1 - \frac{v^2}{c^2}} \times \Delta t_0 \tag{8.7}$$

where v is the vehicle velocity and c is the speed of light. It follows that, as v increases, the passage of time on the vehicle slows down. However, this effect only becomes significant as v becomes close to c. For example, $\Delta t_0 / \Delta t_s = 2.3$, 7.1 and 22.4 for $v = 0.9$, 0.99 and $0.999c$, respectively.

In order to illustrate the potential of relativistic time dilation, let us follow an example described by Sagan (1963). Consider a vehicle capable of uniform acceleration, a, for the first half of its journey and deceleration at the same rate thereafter. Let s be the total distance travelled. The relativistic expression giving the elapsed ship time as a function of a (measured in the ship frame) and s has been given by Sänger (1957):

$$t_s = \frac{2c}{a} \cosh^{-1} \left(1 + \frac{as}{2c^2} \right) \tag{8.8}$$

Following Sagan, fig. (8.1) shows the result of solving eq. (8.8) for $a = 10$ m s^{-2} (i.e. for an acceleration equivalent to one Earth gravity). Classically, a vehicle would be moving at about the speed of light after one year's acceleration at 10 m s^{-2}, so it is clear that relativistic effects dominate within the first year. The distances to some well-known astronomical objects are indicated in fig. (8.1), and it will be seen that the galactic center would be reached in about 20 yr ship time, the Andromeda galaxy in under 30 yr and the quasar 3C 273 in about 40 yr. In fact, such a vehicle would reach the edge of the observable universe in only 44 yr as measured by its crew! For the longer journeys considered here, it may still be desirable to employ some form of suspended animation to supplement the effect of time dilation. Of course, the number of years which would pass on Earth while these trips are in progress is approximately equal to the distance travelled in light-years.

The Interstellar Ramjet

Continuous acceleration for a significant time is difficult to achieve with a rocket because of the very large mass ratios required. Even using antimatter, (and ignoring the price!) continuous acceleration for decades at anything like 10 m s^{-2} would be impossible in practice. In 1960 Bussard proposed a solution whereby an interstellar vehicle would collect its fuel from the interstellar medium using a large scoop. Interstellar protons would then be fed into a fusion reactor and used to produce thrust.

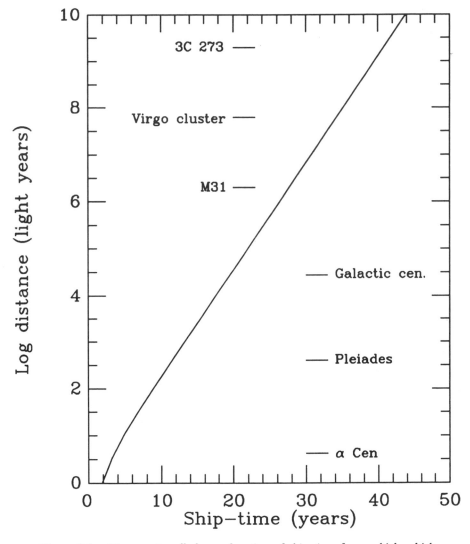

Figure 8.1. Distance travelled as a function of ship-time for a vehicle which accelerates at 10 m s^{-2} for the first half of its journey, and decelerates at the same rate thereafter.

Bussard (1960) showed that the acceleration, a_s, of such a vehicle would be given by

$$a_s = \frac{n m_p c^2 A}{M_s} \alpha \eta \qquad (8.9)$$

where n is the number density of interstellar protons of mass m_p, A is the area of the scoop, M_s is the mass of the vehicle, α is the fraction of the collected mass that is converted into energy, and η is the efficiency with which that energy is converted into exhaust kinetic energy. For a vehicle of mass 3500 t operating in the solar neighborhood, where $n \sim 0.1$ cm^{-3}, eq. (8.9) implies a scoop intake radius of 10^4 km for $a_s = 10$ m s^{-2}. Such a scoop could not be made of solid materials, and Bussard suggested that it be formed from an electric or magnetic field.

The use of electromagnetic fields would require the interstellar medium to be ionized, which is not generally the case. Previous studies have assumed that the local interstellar medium is neutral, and that it would have to be artificially ionized from the vehicle prior to collection. There is now some evidence that the ionization fraction in the local interstellar medium may be quite large (Cox & Reynolds, 1987; Lallement *et al.*, 1994), although even if the local interstellar medium turns out to be sufficiently ionized, there are a number of other reasons why the performance indicated by eq. (8.9) is overoptimistic. First, as noted by Bussard (1960), the p–p chain is extremely slow, requiring the attainment of very high densities in the reactor if it is to work effectively. However, at high density radiation losses from the plasma will exceed the rate of energy production, and a self-sustaining p–p fusion reaction will not be possible (Martin, 1973).

Moreover, there are serious problems with a magnetic scoop, which have been discussed in detail by Fishback (1969) and Martin (1973). First, the necessary field geometry (strongest at the reactor entrance) resembles a 'magnetic mirror', and many incoming particles will actually be reflected by the field. Second, the particles will gyrate about the field lines, and if they do so with radii larger than the physical radius of the entrance aperture, they will not enter the vehicle. In order to maximize the fraction of incident particles that enter the reactor, Martin (1973) found that the magnetic field strength at the entrance to the reactor must increase as a function of velocity: rising from a few hundred tesla at $v = 10^{-4}\ c$ to $\sim 10^7$ T at $v = 0.9c$. As magnetic field strengths of millions of tesla may not be possible from an engineering point of view, Martin considered the effect of a 'technological limit' of 1000 T, and found a maximum intake fraction of only 1.3×10^{-12}. Clearly, a low intake fraction means that the collecting area has to be increased proportionately. Moreover, even if sufficient fuel can be collected, Fishback (1969) has shown that, owing to the finite strength of materials, an interstellar ramjet will be unable to accelerate at 10 m s^{-2} indefinitely, and has produced a modified version of Figure 8.1 which illustrates the effect of this limitation.

Some of the problems associated with a magnetic scoop might be overcome by use of electrostatic fields. Possible configurations of electrostatic scoops have been discussed by Matloff & Fennelly (1977), who point out that, for an interstellar ion density of 10^5 m^{-3}, a 20 C charge would have an effective radius of influence of abut 2.4×10^5 km. Such a charge could be used to attract ions to the vicinity of the spacecraft, where additional electric or magnetic fields would be used to channel them into the reactor.

The many problems with the 'conventional' ramjet concept have led to a number of suggestions to improve its performance, described below.

The Catalytic Ramjet

Whitmire (1975) has suggested that the problems arising from the low cross-section of the p–p reaction may be overcome if the ramjet carries a catalyst to assist proton burning. For example, carbon could be used to enable the reactor to utilize the CNO cycle (such as occurs in hot stars), and other catalytic reactions may be possible. The CNO cycle is many orders of magnitude faster than the p–p chain, and Whitmire was able to show that a catalytic proton-burning ramjet is theoretically capable of maintaining an acceleration of 10 m s^{-2}.

The Ram Augmented Interstellar Rocket (RAIR)

This concept was mentioned in passing by Bussard (1960) and developed by Bond (1974). The idea is for the vehicle to carry its own source of energy, but for it to collect interstellar gas to use as a reaction fluid. Thus it would not have to rely on the p–p chain, but would still be able to collect most of the mass needed for propulsion from the interstellar medium. Of course, acceleration would only be possible for as long as the fuel lasts.

Bond (1974) found that the RAIR reduces the mass ratios required to achieve semirelativistic velocities (0.5–0.7c) by several orders of magnitude relative to those of a conventional rocket. For example, for an ideal RAIR operating with a mass–energy efficiency of $\alpha = 0.002$ (i.e. appropriate for the nuclear burning of hydrogen and lithium) Bond found that a mass ratio of 1.2×10^2 is required to achieve a velocity of 0.7c, compared with a mass ratio of 6.4×10^4 for a conventional rocket using the same reaction. Use of antimatter as the fuel reduces the RAIR mass ratio even further, and has been considered in more detail by Jackson (1980).

The RAIR concept still suffers from the problems inherent in the design and construction of the scoop for the interstellar protons. Bond found that an intake radius of several thousand kilometers is required for operation in a region with 1 proton cm^{-3}, and this would have to be increased by a factor of about 3 for operation in the solar neighborhood (ignoring the scoop inefficiencies discussed by Fishback, 1969, and Martin, 1973, which actually make matters much worse). Thus, although the RAIR offers a great reduction in mass ratios over conventional rockets, this method may not be able to achieve the extreme relativistic velocities needed to exploit time dilation.

The Laser Ramjet

Whitmire & Jackson (1977) have discussed a vehicle which collects its reaction mass from the interstellar medium, but which derives its energy from a solar system based laser. It is therefore similar to the laser powered rocket discussed above, but does not carry its reaction mass. Whitmire & Jackson found that such a vehicle makes more efficient use of the laser beam energy than laser pushed light sails (as does the laser powered rocket), and that, although the conventional ramjet is more energy-efficient at velocities greater than about $0.14c$, this variant does avoid the need for p–p fusion. However, owing to the Doppler effect, it will not be possible for such a vehicle to operate at the relativistic velocities needed for time dilation.

The Ramjet Runway

Whitmire & Jackson (1977) have suggested the possibility of laying down a track of deuterium pellets for later collection by a ramjet. The pellets would be electro-magnetically accelerated from the solar system along the route later to be followed by the main vehicle. Provided that accurate collimation can be maintained, this concept would greatly ease the problems associated with large magnetic ram-scoops. However, it is not clear that sufficient fuel could be deposited along the runway to achieve a large acceleration, and, in any case, acceleration would cease once the vehicle reached the end of the runway. Nevertheless, the concept may be of value for missions which require only semirelativistic velocities ($\sim 0.1c$), and its application to such missions has been discussed by Matloff (1979).

Conclusion

It will be clear from the above that neither the technology nor the economic base necessary to achieve rapid interstellar flight exists at present, and will not exist for decades or centuries to come. On the other hand, it is also clear that interstellar travel violates no physical law, and is therefore a legitimate technological goal for the distant future. We have already achieved the capability of very slow interstellar travel, and there is no reason to believe that the much higher velocities considered here will forever be out of reach. Indeed, nuclear fusion propelled vehicles such as Daedalus may be *technically* possible within the next several decades, although it seems certain that economic and political factors will hold up their development until long after this (Parkinson, 1974; Crawford, 1993b).

Of course, it may be that when interstellar travel finally becomes a reality, it will be achieved by methods quite different from those sketched here. The whole history of technological development leads us to expect the unexpected. The important point, however, is that we can already identify technological solutions to the problem of interstellar travel that are consistent with the laws of physics as we currently understand them. We do not *need* new physics (such as would permit faster-than-light travel, for example), even if such new physics is one day discovered and interstellar travel turns out to be easier than we presently suppose. This conclusion has profound implications for the SETI debate. Attempted solutions to the 'where are they?' paradox which rely on the alleged impossibility of interstellar spaceflight are seen to be untenable. The absence of evidence for extraterrestrials on Planet Earth must therefore be due either to some factor falling within the class of 'sociological' explanations identified by Hart (1975) or, as he argues, to the extreme rarity of technological civilizations in the Galaxy.

Acknowledgements

This chapter is an abridged and revised version of a paper originally published in the *Quarterly Journal of the Royal Astronomical Society*, 31, 377–400 (1990) under the title 'Interstellar travel: A review for astronomers'. I am grateful to the Royal Astronomical Society, and to Blackwell Scientific Publications for permission to reproduce large sections of that article here.

References

BALDIN, A. A. *et al.*, (1988). *JETP Lett.*, **48**, 137.

BENEDIKT, E. T. (1961). *Adv. Astronaut. Sci.*, **6**, 571.

BOND, A. (1974). *J. Brit. Interplanetary Soc.*, **27**, 674.

BOND, A. & MARTIN, A. R. (1978a). In *Project Daedalus—Final Report, J. Brit. Interplanetary Soc. Suppl.*, ed. A. R. Martin, p. S63.

BOND, A. & MARTIN, A. R., (1978b). In *Project Daedalus—Final Report, J. Brit. Interplanetary Soc. Suppl.*, ed. A. R. Martin, p. S37.

BOND, A. & MARTIN, A. R. (1984). *J. Brit. Interplanetary Soc.*, **37**, 254.

BOND, A. & MARTIN, A. R. (1986). *J. Brit. Interplanetary Soc.*, **39**, 385.

BUSSARD, R. W. (1960). *Astronautica Acta*, **6**, 179.

CASSENTI, B. N. (1982). *J. Brit. Interplanetary Soc.*, **35**, 396.

CESARONE, R. J., SERGEYEVSKY, A. B. & KERRIDGE, S. J. (1984). *J. Brit. Interplanetary Soc.*, **36**, 99.

COX, D. P. & REYNOLDS, R. J. (1987). *Ann. Rev. Astron. Astrophys.*, **25**, 303.

CRAWFORD, I. A. (1993a). *J. Brit. Interplanetary Soc.*, **46**, 415.

CRAWFORD, I. A. (1993b). *Spaceflight*, **35**, 188.

CRICK, F. (1981). *Life Itself.* New York: Simon & Schuster.

DIPPREY, D. F. (1975). Appendix to *Frontiers in Propulsion Research*, ed. D. D. Papailiou, JPL TM 33-722.

FINNEY, B. R. & JONES, E. M. (eds.) (1985). *Interstellar Migration and the Human Experience*, Berkeley: University of California Press.

FISHBACK, J. F. (1969). *Astronautica Acta*, **15**, 25.

FORWARD, R. L. (1962). *Science Digest*, **52**, 70.

FORWARD, R. L. (1982). *J. Brit. Interplanetary Soc.*, **35**, 391.

FORWARD, R. L. (1984). *J. Spacecraft*, **21**, 187.

FORWARD, R. L. (1986). *J. Brit. Interplanetary Soc.*, **39**, 328 (correspondence).

FUKUYAMA, F. (1989). *The National Interest*, No. 16, 3 (Summer 1989).

GOLDMAN, T. (1988). In *Antiproton Science and Technology*, ed. R. W. Augenstein *et al.*, p. 123. Singapore: World Scientific.

HANDS, J. (1985). *J. Brit. Interplanetary Soc.*, **38**, 139.

HART, M. H. (1975). *Quart. J. Royal Astronomical Soc.*, **16**, 128.

HOLMES, D. L. (1984). *J. Brit. Interplanetary Soc.*, **37**, 296.

JACKSON, A. A. (1980). *J. Brit. Interplanetary Soc.*, **33**, 117.

JACKSON, A. A. & WHITMIRE, D. P. (1978). *J. Brit. Interplanetary Soc.*, **31**, 335.

LALLEMENT, R. *et al.* (1994). *Astron. Astrophys.*, **286**, 898.

LARSON, D. J. (1988). In *Antiproton Science and Technology*, ed. R. W. Augenstein *et al.*, p. 202. Singapore: World Scientific.

MALLOVE, E. F., FORWARD, R. L., PAPROTNY, Z. & LEHMANN, J. (1980). *J. Brit. Interplanetary Soc.*, **33**, 201.

MARTIN, A. R. (1973). *Astronautica Acta*, **18**, 1.

MARTIN, A. R. (1978). In *Project Daedalus – Final Report, J. Brit. Interplanetary Soc. Suppl.*, ed. A. R. Martin; p. S116.

MARTIN, A. R. (1984). *J. Brit. Interplanetary Soc.*, **37**, 243.

MARTIN, A. R. & BOND, A. (1978). In *Project Daedalus – Final Report, J. Brit. Interplanetary Soc. Suppl.*, ed. A. R. Martin; p. S44.

MARTIN, A. R. & BOND, A. (1979). *J. Brit. Interplanetary Soc.*, **32**, 283.

MARX, G. (1966). *Nature*, **211**, 22.

MATLOFF, G. L. (1979). *J. Brit. Interplanetary Soc.*, **32**, 219.

MATLOFF, G. L. & FENNELLY, A. J. (1977). *J. Brit. Interplanetary Soc.*, **30**, 213.

MICHAELIS, M. M. & BINGHAM, R. (1988). *Laser and Particle Beams*, **6**, 83.

MITCHELL, J. B. A. (1988). In *Antiproton Science and Technology*, ed. R. W. Augenstein *et al.*, p. 359. Singapore: World Scientific.

MOECKEL, W. E. (1972). *J. Spacecraft*, **9**, 942.

MORGAN, D. L. (1982). *J. Brit. Interplanetary Soc.*, **35**, 405.

MORGAN, D. L. (1988). In *Antiproton Science and Technology*, ed. R. W. Augenstein *et al.*, p. 530. Singapore: World Scientific.

PAPROTNY, Z., LEHMANN, J. & PRYTZ, J. (1984). *J. Brit. Interplanetary Soc.*, **37**, 502.

PAPROTNY, Z., LEHMANN, J. & PRYTZ, J. (1986). *J. Brit. Interplanetary Soc.*, **39**, 127.

PAPROTNY, Z., LEHMANN, J. & PRYTZ, J. (1987). *J. Brit. Interplanetary Soc.*, **40**, 353.

PARKINSON, R. C. (1974). *J. Brit. Interplanetary Soc.*, **27**, 692.

PARKINSON, R. C. (1978). In *Project Daedalus – Final Report, J. Brit. Interplanetary Soc. Suppl.*, ed. A. R. Martin, p. S83.

POWELL, C. (1975). *J. Brit. Interplanetary Soc.*, **28**, 546.

REDDING, J. L. (1967). *Nature*, **213**, 588.

REGIS, E. (1985). In *Interstellar Migration and the Human Experience*, ed. B. R. Finney & E. M. Jones, p. 248. Berkeley: University of California Press.

SAGAN, C. (1963). *Planet. Space Sci.*, **11**, 485.

SÄNGER, E. (1957). *Astronautica Acta*, **3**, 89.

SHEPHERD, L. R. (1952). *J. Brit. Interplanetary Soc.*, **11**, 149.

SINGER, C. E. (1980). *J. Brit. Interplanetary Soc.*, **33**, 107.

STRONG, J. & BOND, A. (1978). In *Project Daedalus – Final Report, J. Brit. Interplanetary Soc. Suppl.*, ed. A. R. Martin, p. S90.

STWALLEY, W. C. (1988). In *Antiproton Science and Technology*, ed. R. W. Augenstein *et al.*, p. 373. Singapore: World Scientific.

VANDERMEULEN, J., (1972). In *Proc. Symp. on Nucleon-antinucleon Annihilations* (CERN publication 72–10, ed. L. Montanet), p. 113.

VON EGIDY, T. (1987). *Nature*, **328**, 773.

VULPETTI, G. (1986). *J. Brit. Interplanetary Soc.*, **39**, 391.

WHITMIRE, D. P. (1975). *Acta Astronautica*, **2**, 497.

WHITMIRE, D. P. & JACKSON, A. A. (1977). *J. Brit. Interplanetary Soc.*, **30**, 223.

9

Settlements in Space, and Interstellar Travel

CLIFF SINGER

Interstellar Travel and Extraterrestrial Intelligence

The success of several proposals to search for extraterrestrial intelligence (ETI) in the Galaxy (Cocconi & Morrison, 1959; Oliver & Billingham, 1971; Michaud, 1979) requires the existence of a large number of technologically competent cultures over a long period of time. For example, to expect to find one ETI within 1000 light-years in a perfectly efficient search would require about a million ETI in the Galaxy, each signalling for a million years. (Or it would require 10^8 ETI signalling 10^4 years, or 10^4 ETI signalling 10^8 years, etc.) Many people have asked why some of these ETI should not have taken advantage of their prolonged technological capability to find a method for interstellar travel and settlement of nearby stellar systems (see, e.g., Hart, 1975; Jones, 1976; Winterberg, 1979). If the initial problem of interstellar travel and settlement were solved, then it should become progressively easier for daughter settlements to eventually continue the process until every available stellar system in the Galaxy (including possibly our own) were inhabited.

The chances of this happening have been discussed extensively, often with minimal thought given to the physical requirements for interstellar settlement. In particular, it has been argued that interstellar settlement is either impossible (see, e.g., Purcell, 1960; Marx, 1973) or absurdly expensive (e.g. requiring trillions of man-years of effort to amass the nuclear fuel needed). These arguments require some attention, especially in the light of the possibility that some of the best-known advanced propulsion methods (e.g. pulsed pure fusion: Bond *et al.*, 1978; Winterberg, 1979) *may* be *intrinsically* unworkable. Given such uncertainties (and more profound ones discussed in Appendix A), it would be naive to expect we can determine the minimum physical requirements for interstellar settlement. However, this does not terminate sensible discussion of this topic. For we can set approximate *maximum* physical requirements by examining the limited set of possibilities potentially available with application of our *present* understanding of physics, biology and engineering. The discussion will, therefore, now be limited to the possibilities available to mankind in this context.

Space Habitats in the Solar System

Some insight into the possibilities for settlement and propulsion in space is available as a result of work at Princeton, NASA-Ames and elsewhere (Johnson & Holbrow, 1977; O'Neill & O'Leary, 1977; Arnold & Duke, 1978; Grey, 1977) on the prospects for space settlement and space manufacturing in Earth orbit in the near future. This work forms a relatively solid base for discussion of at least some of what we may do in the solar system in the next few hundred years. Since discussing our potential for interstellar travel only makes sense with an understanding of the potential scale of space manufacturing activities, opportunities for space manufacturing will be reviewed before turning to interstellar settlement.

Consider first the immediate future. Manned activities in space on a regular basis began with the space shuttle. Unless development capital is significantly curtailed, it is reasonably likely that construction of a small permanent manned orbital laboratory will begin by early next century to support existing activities such as reconnaissance, communications and research.

It is instructive to try to estimate the cost of these space activities in units which relate directly to the human effort involved, in order that factors such as inflation and changing types and efficiency of manufacturing processes will not fundamentally alter the intuitive meaning of the units used. A useful unit is 'millions of person-centuries' (mpc), where one person-century is 200 000 h of human labor. Roughly 0.1 mpc have been expended on space activities over the 30 yr leading to deployment of the space shuttle. Given the importance currently attached to the military, economic and scientific activities discussed above, and the capital invested and the relative efficiency of the shuttle as a launcher, this rate of expenditure is unlikely to decrease by a *large* factor in the foreseeable future.

Discussion of possibilities for further human activities in space centers on industrial processes such as solar power satellite stations, small-scale production of special materials or biological products, and large-scale processing of lunar or asteroidal resources. An important factor concerning human activity in space is that, in the long term, there is considerable motivation to turn to use of extraterrestrial sources for materials to build life support systems for the people involved in space operations with continued human interaction. The motivation is the potential for orders of magnitude reduction in transportation costs for acquiring materials from the Moon (or Earth-approaching asteroids) compared with launching from the surface of the Earth. For very extended or large-scale space activities, the impact on Earth resources and environment could also be minimized by restricting most transportation and manufacturing to outside the Earth's atmosphere.

It is, therefore, appropriate that a revival of interest in constructing habitats in space followed the suggestion by O'Neill and coworkers that bags of soil could be launched from the Moon and used to construct a self-sufficient ecosystem at a 'moderate' cost. Since elaborations of this proposal have been extensively described elsewhere, only features particularly relevant to the problems of inter-stellar settlement will be reviewed here. Topics of particular interest are propulsion, ecosystem design (including mass per person, radiation shielding, leakage rates and minimum size), cost and schedule.

The main propulsion requirement in O'Neill's proposal was for moving the main mass of structural material for the space habitat to its construction site in high Earth orbit. The solution was to place a linear synchronous motor (electromagnetic mass driver) at an appropriate elevated point on the lunar equator. Bags of lunar soil launched from the mass driver at 2.4 km s^{-1} would have to be very accurately aimed to reach a catcher device placed beyond the Moon. This could be accomplished by ensuring that the payload goes through a small aperture about 150 km downrange, across a lunar valley. This accuracy is thought to be readily achievable with a set of progressively more distant course correction devices which charge the payload, measure its position, deflect it electrostatically and then discharge it. In fact, with a 150 km baseline for course correction, it should be possible to launch the payload much more accurately than one can compute the solar and lunar and other perturbations on its orbit. Since these perturbations are very small, it would be possible to use a catcher only a few meters across (cf. articles in O'Neill & O'Leary, 1977).

The first comprehensive ecosystem design for a habitat for 10 000 people (Johnson & Holbrow, 1977) used 74 t per person, the majority of which was for soil and for the container. An additional kilotonne per person of industrial slag was allocated for radiation shielding, largely to protect against solar cosmic rays. Various aspects of the ecosystem recycling problem were analyzed, and it was concluded that it should be possible to design a system with 100% recycling which could retain its atmosphere for centuries. The minimum size of a closed human ecosystem (cf. Gilligan, 1975) was not directly addressed in these studies, because the space habitats were sized for the manufacturing capability. In particular, about 10 000 inhabitants were required in a space manufacturing facility so that in 10 years they could build several 5 GW power satellites or build the structure for a second habitat for 10 000 people. It seems likely from the results of these studies that it would be *possible* to build a smaller *self-contained* ecosystem with 50–500 people with 10 000–100 000 t of structure and biological mass. (It might be cheaper to supplement such a habitat with materials from Earth in the *initial* stages of space settlement, but we are concerned here with a different situation where it is only necessary to scale down existing larger ecosystems.)

The cost of building a manufacturing facility with 10 000 occupants in 22 years has been calculated to be 2×10^{11} 1975 United States dollars, or about 0.2 mpc. While this may be optimistic, the cost estimate is sufficiently detailed so that it is unlikely to be off by an order of magnitude, which is sufficient for the present discussion.

A further development in space manufacturing would be retrieval of small Earth-approaching asteroids (Arnold & Duke, 1978; O'Leary, 1978). Some of these contain a better mix of materials for building ecosystems than lunar soil does. Others could serve as a vast source of metals for space or even terrestrial manufacturing. Estimates of the cost of transporting the best candidates to Earth orbit are already competitive with the lunar alternative. With a better survey of some of the estimated 100 000 Earth-approaching asteroids of appropriate size, and with possible application of very efficient retrieval methods (Singer, 1979), such asteroids may become the initial as well as eventual source of materials for large space habitats.

The rapid construction proposed in the studies described above was timed to provide large supplies of electricity from solar power satellites in the near future. If this does not occur, the habitation of space will probably proceed at a slower pace determined initially by servicing of reconnaissance, communications, research and (Driggers, 1979) specialized manufacturing processes. Eventually, the lower average cost of sustaining such activities should dictate use of closed ecosystems supplemented with extraterrestrial resources, but this could take many decades. A key factor in the habitation of space is the production of one space manufacturing facility capable of building most of a second such facility, but it is difficult to predict whether this will take as little as 20 or as much as 200 years.

To complete the discussion of settlement of the solar system, we note that designs of space settlements either for 100 000 or for over a million people have been outlined (see, e.g., O'Neill, 1974, and references quoted above). While the living conditions in a smaller space habitat are meant to be comparable to those in an affluent city or existing small village, these larger habitats may include open spaces with a scale of many kilometers and possibly even weather. Whether large or small habitats are chosen, it seems clear that the possibility exists to have millions of people living in many structures in space. In the author's opinion, this is remotely possible within something over 50 years, probable within 500 years, and difficult to avoid within less than 5000 years (barring profound biological transformation, incessantly repeated global catastrophes, or the remarkably difficult (Viewing and Horswell, 1978) achievement of catastrophe leading to human extinction).

'Known' Solution to Biological Problems of Interstellar Settlement

Since interstellar transport would very likely require large construction activities in space, the above discussion has concentrated on space settlement, and ignored additional possibilities concerning settlement of planets and satellites. (These intriguing additional possibilities are discussed by J. Oberg in chapter 10; they, too, would probably require substantial habitation in space settlements as a pre-requisite.) The discussion will now concentrate on what would be required for some of the inhabitants of the solar system to move to a nearby star. First the biological problems of the spaceship design and then the propulsion problem will be discussed.

Staying within the context of the present discussion, the only known biological solution is to use a pre-existing closed ecosystem as the interstellar spaceship (i.e. an 'interstellar ark': Matloff, 1976). The fundamental environmental design problems for this solution have been extensively addressed in space habitation studies, as discussed above. Other possible solutions are too speculative to help give any confidence in estimates of maximum requirements for interstellar settle-ment, so discussion of these is relegated to Appendix A. Moreover, the technologi-cal base required for presently known solutions to the propulsion problem cited below already requires extensive experience at designing space ecosystems and inhabiting them for at least a century or more. Therefore, discussion is restricted to a few biological problems which could be unique to this method of interstellar settlement.

One suggested biological problem is in-breeding due to a small genetic pool. This is clearly a 'red herring' for two reasons. First, proper crew selection should introduce great genetic diversity (making the dubious assumption that this is indeed essential). Second, the already widespread practice of storing and transfer-ring human zygotes could provide unlimited diversity with a few grams of such cells.

Another problem is radiation exposure from galactic cosmic rays. In the space habitat studies it was necessary to provide significant additional shielding mass, partly for protection against some galactic cosmic rays, but primarily for protection against solar cosmic rays. Since active shielding against such radiation cannot necessarily be provided during the interstellar voyage, one solution is to accept the order of magnitude increase in payload mass required by passive shielding. Another possibility is to accept the $5-20$ rem yr^{-1} (depending on detailed design) dose without massive shielding (compared with a minimum of 50 rem yr^{-1} quoted as a lower limit required for detectable damage from steady radiation: Johnson & Holbrow, 1977). Crew selection might avoid rare (but perhaps presently epidemiologically significant) radiation-sensitive individuals. Extra shielding for zygotes, and possibly for developing organisms, might also be useful. Since the possible hazards from failure of energy supply (e.g. a several megawatt nuclear reactor), environmental systems and, especially, propulsion systems are likely greatly to outweigh the radiation hazard, it is likely that multiple redundancy in these systems would enhance the probability of success much more than passive shielding.

Finally, there is the question of social stability in a small remote habitat. The following observations may be relevant to this question. First, there is *no* known universal tendency for small isolated human communities to inevitably self-destruct. Second, the possibility of continuous habitation of space habitats in the solar system for centuries could allow extensive opportunities to test social stability under related physical conditions. Third, the most promising propulsion method discussed below could be facilitated by extended human support in the local interstellar medium. This would allow a further testbed for social stability in interstellar arks, should it be necessary. All of the biological and social problems mentioned above look relatively tractable in comparison with the propulsion requirements, which are now discussed.

An Example of Propulsion for Human Interstellar Settlement

Until recently, the best interstellar propulsion method for which reasonable confidence of feasibility can be claimed was the pulsed thermonuclear bomb and pusher plate concept outlined by Dyson (1968). Two other concepts appear interesting but have potentially unsolvable engineering difficulties. These are microfusion explosions ignited by relativistic electron beams (Bond et al. 1978), and use of a highly collimated laser beam pointing out from the solar system (Marx, 1966; Jackson & Whitmire, 1978). The microfusion concept is plagued by uncertainties that plasma stability will allow a net energy gain (much less the high efficiency required for interstellar propulsion). Another problem is the damage potential of the enormous number of neutrons produced in any reasonable microfusion scheme; it is quite likely that detailed results will show that it is not possible to guarantee reabsorption of nearly all of these neutrons during the complete sequence of an efficient microexplosion. While beaming laser energy to an interstellar spaceship would reduce onboard propulsion requirements dramatically, it is far from clear that the required laser optics (e.g. sources about 100 km square with a tolerance below one part in 10^{11}: Singer 1980) are achievable. (The requirements for using laser power for deceleration are even more formidable, although alternative solutions for deceleration may exist, cf. F. Dyson in chapter 7.)

Fortunately, it has recently been possible to look in some detail at another propulsion method which combines the advantages of those discussed above: these are off-board propulsion, relatively high efficiency and reasonable confidence of engineering feasibility. This propulsion method uses an electromagnetic or centrifugal relay mass driver to launch a stream of small pellets using a local solar or nuclear power source (Singer, 1980; Singer & Singer, 1991). An analysis of this method was stimulated by work on mass drivers at Princeton (see, e.g., O'Neill & Snow, 1979), MIT (see, e.g., Kolm et al., 1979) and the Lawrence Livermore Laboratory (Brittingham & Hawke, 1979); by the achievement of acceleration of 360 000 gravities on a 5 m rail gun (Rasleigh & Marshall, 1978); and by investigation of interstellar propulsion problems carried out by the Daedalus team (Bond et al., 1978). It was also preceded by proposals for hypervelocity accelerators by Winterberg (1966), for local applications by Ruppe (1966) and on the potential of mass drivers by Clarke (1950). Another critical observation was the suggestion that projectiles could be aimed extremely accurately by successive course corrections over ever-longer baselines, as discussed above for the Moon-based launcher. Relatively pessimistic assumptions about interstellar dispersion of a collimated pellet stream show that a modest number of course correction devices would be required to retain collimation of the stream over distances of light-years. Finally, the pusher plate concepts used in the thermonuclear bomb and microfusion schemes were readily adapted to the pellet stream concept, with the advantage that ionization of the pellets near the pusher plate would give a clean low-temperature plasma with negligible neutron flux.

One of many possible scenarios using pellet stream propulsion has been analyzed to provide a concrete example of an interstellar settlement mission. The sequence of events in this mission is illustrated schematically in fig. 9.1. The events are as follows.

1. Preparation for this mission begins 110 years before the departure of the settlers with the launch of two slow pellet streams from a relatively short mass driver. (The mass driver is shown with its length outlined against the Sun in the upper left of fig. 9.1.) These slow pellet streams will be used for partial deceleration of the payloads before they reach the target, Proxima Centauri. The upper, faster stream will be used by a small lead ship, and the lower, slower stream by the main body of settlers.
2. For 89 years, faster pellets traveling at 39 000 km s^{-1} are intercepted by the main ship. The pellets are launched from a mass driver consisting of 100 000 segments each 3000 m long.
3. For another 13 years, another stream of similar pellets continues to accelerate the lead ship.

PROFILE OF MANNED MISSION

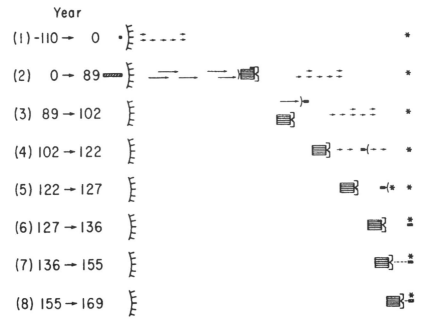

Figure 9.1   Schematic illustrations of steps in an interstellar settlement mission. Arrows on pellets indicate velocities of the various pellet streams used for acceleration and partial deceleration.

4. First the lead ship and then the main ship partly decelerate by running into the slow pellet streams.
5. The lead ship eventually stops, using fusion bombs and a pusher plate for deceleration (an alternative deceleration method is outlined in Appendix B). Thirty-eight years after separating from the main ship, the lead ship arrives at Proxima Centauri and
6. begins construction of a third miniature pellet launcher.
7. This launcher will lay out a very slow stream of pellets (with maximum velocity of 610 km s^{-1} toward the main ship) to be used for
8. final deceleration of the main ship.

The settlers could then use asteroidal planetary material at the destination for expanding their living space or building new habitats.

A summary of the mission requirements is given in table 9.1. (The physical parameters for table 9.1 and fig. 9.1 were computed using the equations of Singer, 1980, and Dyson, 1968.) Guesses at the cost of the major components (based on

Table 9.1. *Nominal mission requirements with 'present' technology*

Component	Size		Cost (mpc)
Launcher			
(a) accel.	10^5 of 3000 m segments		0.3
(b) decel.	2000 of 3000 m segments		0.01
Power supply	40 TW maximum		0.1
	Main vessel	*Lead ship*	
Ship mass	100 kt	100 kt[a]	0.01
Thermonuclear fuel	——	1000 kt[a]	0.3
Settlers	1000	100	0.002
		Total =	0.72

[a]Most of the mass of the lead ship and its thermonuclear fuel is delivered in another pellet stream (not shown in fig. 9.1). The velocity of the lead ship after acceleration is adjusted to approximately match the velocity of this supply stream.

guesses from Singer, 1980, and Dyson, 1968) are also listed in table 9.1. The total additional effort required for space habitats established in the solar system to launch an interstellar settlement is estimated at just under 1 mpc. The major uncertainty in this analysis is the efficiency of the downrange segments of the largest mass driver (one order of magnitude?). A smaller allowed specific power in the pellet reflector and/or the efficiency of the pulsed thermonuclear deceleration could also increase the mission cost by as much as an order of magnitude. Should such limitations be encountered, it is likely that a longer mission time would be chosen. Even a factor of 2 or 3 increase in the mission time t_* would allow of a large reduction in the required capital investment. For example, the required number of mass driver segments scales as t_*^{-2} (and they may be easy to build when lower pellet velocities are required). And the difficulty of handling the high power density incident on the spaceship's reflector plate decreases at least as t_*^{-3}. Thus, it seems likely that settlement missions requiring 170–500 years of travel time could be achieved with a total investment within an order of magnitude of 1 mpc.

Conclusions

The above discussion can be summarized very succinctly as follows. We can be reasonably confident that it should be physically possible to begin to send settlements to nearby stellar systems after establishing a population of many millions in space habitats in the solar system. This is likely to happen several centuries after the invention of the radio telescope, and could conceivably require only one century. The time required to set up such habitats with a modest fraction of the available manpower is in any case certainly much less than 10^4 years. A maximum limit of 2–5 centuries on the time required to transport a self-contained human ecosystem to the nearest stars can be inferred from our present understanding of the biological and physical problems involved. Establishing the capability to launch such missions should be possible with an investment of the order of a million person-centuries.

Now consider the implications of these conclusions for the argument that interstellar settlement is physically unrealistic for cultures which are potential sources of communication from ETI. Since the technological lifetimes of these cultures exceeds 10^4 years in even the most extreme scenarios supporting ETI search strategies, it appears that every source of communication from ETI in the Galaxy has had ample opportunity to instigate interstellar settlement. Whether any of them would want to has been questioned. The only comment made here is the following. If ETI have resources comparable with our own, then success of an ambitious ETI search would imply that the equivalent of at least 10^{15} mpc had existed in the Galaxy prior to our receiving the first communication. The comparison of this number with the order of 1 mpc required for us to initiate interstellar settlement makes an impressive contrast.

Although a careful analysis of the full implications of these results is beyond the scope of this chapter, one possible conclusion is evident: that it may be extremely difficult or impossible to find sources of communication from ETI in the Galaxy. In fact, in the light of a sober reappraisal of our minute but increasing understanding of the large variety of possible factors in astrophysical and chemical dynamics which may be required for the appearance of technological intelligence, it would be far from surprising if we were the only technological culture in the Galaxy. In this case, in the words of a reporter who covered the conference on which this book is based, 'perhaps it is the great destiny of man himself to spread life through the Galaxy'.

Hopefully, it will be obvious that the important question is not how quickly we can fulfil such a destiny, but rather what the quality of the life we spread through the Galaxy will be. These may be the sentiments expressed in this fragment of a poem by Richard Brautigan (1977):

I like this planet.
It's my home and I think it needs our attention and our love.
Let the stars wait a little while longer.
They are good at it.
We'll join them soon enough.
We'll be there.

Appendix A: Alternative Modes of Biological Transport

The discussion of interstellar settlement in the main text was restricted to consideration of small human communities reproducing by methods already considered acceptable. This assumption is useful for studying the near future or for constructing an 'existence proof' of the possibility of interstellar settlement. But it is far from clear that it is safe to assume that the same biological problems will be faced by ETI or even by our own progeny in the distant future. Given that recent natural biological evolution has produced profound changes in the form of intelligent life on Earth in a minute fraction of galactic history, there would certainly be further evolution even without human intervention.

However, it is becoming increasingly clear that over the next few decades and centuries, we will acquire increasing control over technology relevant to genetics. At the least, this will probably involve decoding the structure and function of a significant part of human and other genomes, and the ability to construct genes of any desired nucleotide sequence and insert them into many kinds of cells. That there will be a concerted attempt to use such technology to effect certain improvements in human genomes (e.g. cure of genetic defects and/or of various diseases) is also already quite clear. It seems plausible to suggest that the range of possible further developments in genetic engineering allowed of by the technological base will be so large that social rather than technological factors will limit what is actually done.

The range of biological possibilities in a galaxy with a large number of long-lived technological ETI (as presumed in most communication search strategies) is evidently enormous. To begin with, it is only a plausible but unprovable hypothesis that *natural* evolution of ETI primarily occurs on timescales and in physical conditions which produce (to within a modest factor) ETI with size and lifetime comparable with ours. Were there large numbers of ETI in the Galaxy, then the possibility of *artificial* control of genetics to produce mature ETI of different size and, particularly, different longevity might well be available to many cultures or subcultures. Moreover, one can imagine different modes of reproduction which produce physically similar mature ETI but which could have profound implications for interstellar settlement.

Only a few of these myriad possibilities will be discussed. In particular, a discussion of the traditional example of hibernation of human communities is avoided. Hibernation would still require some life support system during transit and might very likely require a sizeable ecosystem after arrival, at least for travel to nearby stars. It might therefore only save a modest factor in the mass to be transported. Moreover, since primates do not hibernate, it might also require significant genetic redesign. Therefore, only more interesting examples of the possibilities with active control of genetics are discussed here. It is certainly not claimed that we or other ETI could or would necessarily want to achieve all of these possibilities. But it is claimed that the existence of something similar to at least some of these alternatives is plausible. The range of biological alternatives makes it extremely difficult to place any lower physical limit on the speed of interstellar settlement which is relevant on the timescale of galactic evolution.

First, consider ETI which have achieved longevity and psychological stability of 1000 years or more. One or two such individuals could conceivably be sent on a settlement journey in a ship orders of magnitude smaller than the large interstellar ark described in the main text. Particularly if large bombs were not necessary for deceleration (cf. Appendix B), this could allow of very small spaceships. The ability to undertake millennial voyages could also greatly relax the propulsion requirements. If the individuals were, say, an order of magnitude smaller than humans, it could reduce the propulsion requirements still further. If necessary, they could presumably carry along a sperm or zygote bank. (Building a radio receiver to receive the latest word on desirable nucleotide sequences for genomes might be an alternative to a zygote bank.)

Alternatively, the zygotes could even be matured at destination by an automatic device, assuming that this would result in lower spaceship mass than carrying a mature ETI. This may seem a highly unpalatable prospect from a human point of view, but it cannot be logically excluded.

Another possibility with some terrestrial analogs is that an ETI might have naturally or intentionally adapted to living in space in slightly more radical ways. A few millimeters of protective film exuded on the surface of an ETI (or a non-sentient biological 'house') might allow direct implantation and growth of a small immature ETI on the surface of an asteroid in space. Should deceleration methods like those described in Appendix B be possible, then an immature ETI with a mass of a gram or less might be launched directly from a long mass driver with relatively modest acceleration in a 'ship' of mass from a gram to a tonne. Contrary to current ideas, it could therefore theoretically be possible for a single

ETI to send self-reproducing 'settlements' at very high velocity to millions or billions of stellar systems, even to neighboring galaxies.

This last suggestion goes far beyond what is necessary to demonstrate the importance of considering biological alternatives when attempting to make definitive statements about the limits to interstellar settlement, and it may in fact be physically impossible. But the less radical examples clearly demonstrate that assuming known biological alternatives is only useful in putting an *upper* limit on the difficulty of interstellar settlement.

Appendix B: Deceleration on Blowback from Colliding Pellets

Deceleration is a major problem for a stellar orbiter probe or settlement mission using any propulsion method. For onboard propulsion, either the mission time is doubled or the fuel requirements are typically greatly increased in comparison with a fly-by mission. Were remote electromagnetic propulsion made practical (e.g. by periodic refocusing of the beam) deceleration would be particularly problematic. Deceleration on a slow pellet stream is promising, but it is limited by velocity of the prelaunched pellets which the decelerating ship runs into.

The alternative of electromagnetic braking on a set of fine wires (with minimum size limited by erosion) is discussed by F. Dyson in chapter 7. Assuming that a localized hyperalfvenic shock forms around each wire, there is a question as to whether the resulting turbulent wake would produce significant drag. (A reliable answer to this question would require experimental support.)

An alternative suggestion is to produce a stationary plasma in front of the decelerating spaceship using colliding pellets. This idea is illustrated in fig. 9.2. Two streams of pellets launched well before the beginning of the mission would be collimated as described in the text. Longitudinal velocity corrections by device(s) released from the spaceship would ensure that pairs of pellets from each stream pass through a collimator on the ship near the same time. A slower, shaped, pellet would be overtaken by a smaller, faster, pellet. The slower pellet would be shaped so that the resulting impact explosion would blow back part of the resulting plasma toward a reflecting plate on the ship. (Some thermonuclear fuel in the shaped pellet might conceivably increase the blowback fraction at relatively low impact velocities, but this refinement is not obviously essential.) Even if the blowback fraction were not large or if the blowback were still traveling slowly toward the target star, this deceleration method could still be a major improvement over other methods such as nuclear bomb explosions. Not only would the need for large quantities of fissile material be avoided, but also the minimum size of the

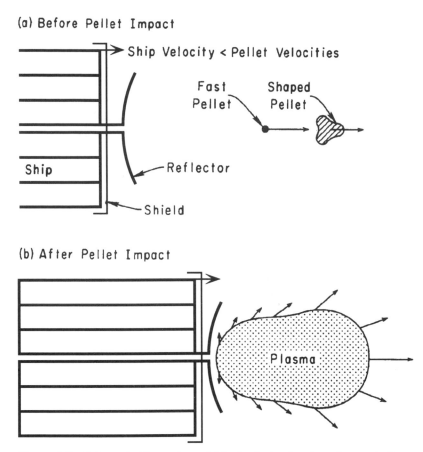

Figure 9.2 Schematic illustration of blowback deceleration. Arrows indicate relative velocities as viewed from Earth. Sizes of ship and pellets are not to scale.

reflector plate could be much smaller and an efficient design might be much easier.

Either electromagnetic deceleration or the blowback technique could significantly reduce the investment required both for the conservative biological approach to interstellar settlement summarized in the main text and for the more radical biological alternatives discussed in Appendix A.

Acknowledgements

I should like to thank H. Taylor for his flexibility in administering a National Science Foundation Postdoctoral Fellowship which supported the germination of these ideas years ago, F. Dyson for pointing out the existence of the work which led to the idea presented in Appendix B, and B. Pieke for reading the manuscript. I also thank E. Carey for typing the manuscript when she could have been drinking tea, and the Princeton Plasma Physics Laboratory for the artwork.

References

ARNOLD, J. R. AND DUKE, M. G. (editors), (1978). *1977 Summer Workshop on Near-Earth Resources*, La Jolla, CA., NASA Conf. Pub. 2031.

BOND, A., MARTIN, A. R., BUCKLAND, R. A., GRANT, T. J., LAWTON, A. T., MATTISON, H. R., PARFITT, J. A., PARKINSON, R. C., RICHARDS, G. R., STRONG, J. G., WEBB, G. M., WHITE, A. G. A., AND WRIGHT, P. P. (1978). *J. Brit. Interplanetary Soc. Suppl.*

BRAUTIGAN, R. (1977). In *Space Colonies* (editor: S. Brand), Penguin London, p. 51).

BRITTINGHAM, J. N., AND HAWKE, R. S. (1979). Devices for Launching 0.1-g Projectiles to 150 km/s or More to Initiate Fusion, UCRL-52778.

CLARKE, A. C. (1950). *J. Brit. Interplanetary Soc.*, 9, 261.

COCCONI, G., AND MORRISON, P. (1959). *Nature*, 184, 844.

DRIGGERS, G. W. (1979). Is Lunar Material Use Practical in a Non-SPS Scenario? Paper 79-1414, *Fourth Princeton/AIAA Conference on Space Manufacturing*, Princeton.

DYSON, F. J. (1968). *Physics Today*, 19, (Number 10), 41.

GILLIGAN, E. S. (1975). *Migration to the Stars* (Luce, Washington).

GREY, J. G. (editor), (1977). *Space Manufacturing Facilities*, Vol. 1 and Vol. 2 (Proc. 1974, 1975, and 1977 Princeton/AIAA Conference on Space Manufacturing Facilities), AIAA: New York.

HART, M. H. (1975). *Quarterly J. Royal Astron. Soc.*, 16, 128.

JACKSON, IV, A. A., AND WHITMIRE, D. P. (1978). *J. Brit. Interplanetary Soc.*, 31, 335.

JOHNSON, R. D., AND HOLBROW, C. (editors), (1977). *Space Settlements: A Design Study*, 1975 Stanford University-AMCS Research Center Summer Faculty Fellowship Program in Engineering Systems Design, NASA SP-413.

JONES, E. M. (1976). *Icarus*, 28, 421.

KOLM, H., FINE, K., MONGEAU, P., AND WILLIAMS, F. (1979). Electromagnetic Propulsion, Paper 79–1400, *Fourth Princeton/AIAA Conference on Space Manufacturing Facilities*, Princeton.

MARX, G., (1966). *Nature*, 211, 22.

MARX, G., (1973). In *Communication with Extraterrestrial Intelligence*, (editor: C. Sagan), MIT Press, Cambridge, Mass. p. 226.

MATLOFF, G. L., (1976). *J. Brit. Interplanetary Soc.*, 29, 775.

MICHAUD, M. A. G., (1979). *J. Brit. Interplanetary Soc.*, 32, 116.

OLIVER, B. M., AND BILLINGHAM, J., (1971). *Project Cyclops: A Design Study of a System for Detecting Extraterrestrial Life*, NASA CR 114445.

O'LEARY, B., (1978). *Astronomy* (number 6), p. 6.

O'NEILL, G. K., (1974). *Physics Today* 27, (number 9), 32.

O'NEILL, G. K., AND O'LEARY, B. (editors), (1977). Progress in Astronautics and Aeronautics, 57, *Space-Based Manufacturing from Non-terrestrial Materials; 1976 NASA-Ames Study*, AIAA: New York.

O'NEILL, G. K., AND SNOW, W. R., (1979). Overview and Outline of Mass-Driver Two, Paper 79–1396, *Fourth Princeton/AIAA Conference on Space Manufacturing Facilities*, Princeton.

PURCELL, E. M., (1960). In *Interstellar Communication*, (editor: A. G. W. Cameron), W. A. Benjamin, New York, p. 121.

RASLEIGH, S. C., AND MARSHALL, R. A., (1978). *J. Appl. Phys.*, 49, 2540.

RUPPE, H. O., (1966). *Introduction to Astronautics and Aeronautics Vol. I.* (Academic, New York), p. 20.

SINGER, C. E., (1979). Collisional Orbital Change of Asteroidal Materials, Paper 79–1434, *Fourth Princeton/AIAA Conference on Space Manufacturing Facilities*, Princeton.

SINGER, C. E., (1980). *J. Brit. Interplanetary Soc.*, 33, 107.

SINGER, C. E. AND SINGER, F. R. (1991). *J. Brit. Interplanetary Soc.*, 44, 127; c.f. Ouzidane, M. (1993). 'Spaceborne Centrifugal Relays for Spacecraft Propulsion, University of Illinois at Urbana-Champaign Dept. Aeronautical and Astronautical Engineering Ph.D. thesis (summary submitted to *Acta Astronautica*, 1994).

VIEWING, D. R. J., AND HORSWELL, C. J. (1978). *J. Brit. Interplanetary Soc.*, 31, 209.

WINTERBERG, F. (1966). *Plasma Phys.*, 8, 541.

WINTERBERG, F. (1979). *J. Brit. Interplanetary Soc.*, 32, 403.

10
Terraforming

JAMES OBERG

Earth is unique in this solar system – it is the only planet that seems to support life. Its hospitable ecosphere stands in stark contrast to the empty, lifeless landscapes of the Moon, Mars, Venus, Mercury and other worlds probed by our spacecraft. Recent arguments suggest that while planets may be common in the universe, habitable worlds may not be. Internally, a candidate planet's proper composition, tectonic dynamics and very narrow but extremely long-period thermal stability may be rare. External biosphere-destroying natural processes, from interstellar dust clouds to sterilizing radiation sources, may periodically rid vast regions of a galaxy of planet-bound life forms.

Such a severe limitation on life-supporting worlds significantly impacts discussions of the Search for Extraterrestrial Intelligence at both ends of the question, the search for causes and the search for consequences. On the former issue, it seriously reduces the stage on which the formation and evolution of life may take its chances against the odds: few candidate planets means even fewer ultimate successes. At the other end of the question, it suggests strategies for searching for the few successful technologies that do evolve, by identifying potential technological activities they would choose to engage in, activities that may have very long range of detectability. The significance to human searches for ETI may be profound.

The issues of 'probability of intelligence arising' are dealt with in other chapters. My purpose is to address the question of final consequences. What kind of technological activity would be attractive to a civilization on a rare habitable planet? Our sample size is 1, which makes us 'average'. But we have the imagination to hypothesize alternatives and variations. What do they suggest?

The hypothesized rarity and brevity of Earth-like worlds may be subject to technological redress. Even if Earth is unique and alone and threatened with future external assault, it doesn't have to stay that way. Future terrestrial tool-users (human or otherwise) may some day be able to increase the number of life-supporting worlds in this solar system from 1, as at present, to a dozen or more. Instead of just being a freak accident of biology, Earth could serve as a blueprint for the transformation of her sterile sister worlds into Earth-like biospheres. There is no need to passively hope and wait for this metamorphosis to come about through billions of years of random, rare accidents. It can be made to happen in just a few centuries of deliberate technological manipulation.

The term for this awesome concept is 'terraforming'. The word was coined 50 years ago in a science fiction story, but the concept of purposeful world shaping goes back much further in human literature, to the creation myths of all cultures. How far it may still be ahead of its time is unknown. But it is not an alien, unthinkable idea.

Advocates of terraforming conjure up idyllic visions of reshaped planets made fit for human settlements. They ask incredulous listeners to imagine red-skyed Mars watered and in bloom; to imagine choking Venus tamed and cooled; to imagine even Earth's sterile, airless Moon transformed into a smaller replica of Earth.

And how could such planetary remodelling be accomplished? The transformations, it turns out, do not really require magic but only extrapolation from what we know today. Colossal energies would be needed, but in a few centuries such forces should be available. Intimate knowledge of climatology would be needed, but those lessons must be learned on Earth in any case.

Manipulations of biology and ecology are still far beyond present scientific capabilities, but they lie along the directions in which modern science and technology are moving.

Each candidate planet would require a different combination of these techniques, although many of the tools would be common from world to world. The major obstacle is not technological but conceptual: Humanity does not yet realize that it has the capability to transform whole planets, not only for worse (as we are often warned by the doomsday prophets), but also for better.

A good first candidate is the planet Venus, once thought to be a twin of Earth but now known to be a closer analog of medieval visions of hell. The planet gets too much sunlight, has too much carbon dioxide and sulfuric acid, and has a day–night cycle far too long.

Imagine what would happen to a human being placed on the surface of Venus. Immense atmospheric pressures would instantly crush the soft body tissues, while oven-like temperatures would convert body water to steam. Layers of charring flesh would peel off explosively. An expanding cloud of soot would surround a pile of crumbling bones as the acidic vapors of Venus turned a human body to dust.

Terraformers call for a physical assault on such hostile conditions. Carbon dioxide and other chemicals in the atmosphere of Venus would be transformed biologically by clouds of algae, suitably tailored in genetics laboratories to thrive under Venusian conditions. Artificial dust clouds or 1000-km-wide parasols would shade the planet, in orbits that mimicked a near-24-h day–night cycle. The excess mass of oxygen in the atmosphere (about 60 bar worth) would have to be physically ejected or – better yet – combined with hydrogen imported from the icy asteroids beyond Jupiter, to form water oceans and limestone.

A century or two after this transformation, human beings would walk in the upland regions of Venus without backpack refrigerators and perhaps even find it pleasant. Along the newly formed coasts of the Aphrodite and Lakshmi continents, the climate could resemble that of Samoa or Curaçao or the Côte d'Azur.

Mars offers different problems. A human body placed on Mars would exhale all of its internal gases in a great rush out of body orifices. Consciousness would fade through anoxia as the thin Martian air provided no breath. The cold of the sand would freeze the fallen body within minutes, but it might take millions of years for the mummified flesh to erode. In the end, one patch of red sand might have a lighter coloration; beyond that, the visitor would leave no trace.

Mars has neither enough air nor enough sunlight, although its day is nearly Earth-like. To attract more sunlight, dark soot could be mined from the two small moonlets in orbit around Mars, and spread on the surface of the planet. If permafrost (giant dust-covered glaciers) exist, it might melt, flooding the surface after a billion years of drought; in cases where enough such native water does not exist, asteroidal water would have to be imported. Biological activities could be instigated, perhaps in miles-deep oasis valleys gouged out of the landscape by the ice melt or by the impact of incoming asteroids (which also would heat these wide 'oasis craters'). Additional heating could be provided by giant space mirrors, 1000 km on a side, concentrating sunlight onto the planet.

As the Martian air thickens to breathable levels, and as temperatures rise to above the freezing point of water, a new climate could be formed that would approximate that of the Andes or the Caucasus or Kashmir. If such conditions on the new planet seem unattractive, recall the persistent stories of longevity and happiness among mountain people here on Earth.

Earth's own Moon need not remain forever barren. Even as human mining activities bring it a measure of life in the coming decades, it, too, could hold an atmosphere, either baked from its own rocks or imported from Saturn and beyond (a 200 km ice cube would provide an adequate atmosphere that lunar gravity could hold for thousands of years). Other rocky worlds such as Mercury and various moons of the outer planets such as Ganymede, Titan and Triton could similarly be engineered into habitable home worlds.

There is a new candidate whose qualifications have only recently been recognized: Io, innermost of the Galilean worlds circling Jupiter. Its sulfur volcanoes blasted their way into human consciousness via the Voyager probes, driving home the lesson that in space one can only dare expect the unexpected. It is in keeping with this still-unfaded sense of wonder, elicited by the Voyager vistas, that I want to nominate Io for terraforming – and at the top of the heap, as well.

Io has several advantages over more classical candidates. First, it has an internal heat source generated by tidal stresses induced by Jupiter (most of the surface is cold because in a vacuum the outflowing heat leaks away quickly). Second, its 42.5 h day is not grossly different from that of Earth. Third, it is deep within the magnetic field of Jupiter, a factor which has biochemical advantages lacking on all other solid planets beyond Earth.

But of course Io has some powerful drawbacks, at present. First is the killing radiation belt which surrounds Jupiter. Second is the lack of water or, for that matter, any atmosphere worth sneezing into. Third is the surface enrichment of sulfur compounds spewed forth from underground lakes of molten sulfur. Each of these problems by itself might seem to veto any consideration of terraforming Io and appear to counterbalance the substantial advantages enumerated above.

Well, maybe not. Radiation belts are swept out by rings of rocky debris – as we learned during Saturn fly-by missions. The jovian radiation in the neighborhood of Io could thus be decontaminated by pulverizing a large fraction of the small inner carbonaceous moon, Amalthea (or the even smaller moons discovered inside Amalthea's orbit), forming an artificial ring extending out to and enveloping Io.

Next, water would have to be imported, either from ice-rich, sibling worlds Ganymede and Callisto or by impacting some outer jovian moons or nearby asteroids onto Io (comets are too small and too unpredictable to be worth chasing). Note that the sulfur, while plentiful on the surface because of differentiation, should not be more abundant in proportion to the whole mass of the planet than it is on Earth. The surface sulfur could be buried using dirt excavated by the impact force of the 'ice-teroids' carrying the components of the future atmosphere and ocean. Subsequently, biological tailoring could begin, lasting many decades – then spacesuits on Io would become obsolete. Io could be made a habitable world by the end of the next century.

This vision is at present unconstrained by harsh reality. Perhaps its thermal state is too active to support a stable crust (then we can turn our attention to cooler, wetter Europa). Perhaps the asteroidal engineering needed to 'short circuit' the jovian radiation belts is too ambitious or unreliable. Perhaps some form of carbon–sulfur life exists in the hot springs or, even more bizarre, swims in the liquid sulfur ocean beneath the crust.

Terraforming studies consider the conditions needed on habitable planets, and the tools and techniques which are conceivable today. These topics fill several chapters in my book *New Earths* (Stackpole, 1981), the first non-fiction treatment of the concept of terraforming. Several subsequent issues of the *Journal of the British Interplanetary Society* have been devoted to evolving concepts. Future lines of inquiry are evident: climatological goals can be quantified; roadblocks can be identified and areas of current ignorance defined; strategies can be suggested. The bottom line is that terraforming could well be possible given enough time and enough money and enough intelligence.

Now turn this futuristic concept toward the problem of SETI. If naturally-occurring Earth-like worlds are truly rare in the universe, and if star faring civilizations (both hypothetical current ETIs, along with our own also hypothetical descendants) retain a desire to live on (or even just vacation on) planetary surfaces, they may not be able to find sufficient hospitable planets. In that case they may have to make their own, and everything so far indicates that such planetary engineering is feasible.

Engineering planetary surfaces would be only the beginning for such an advanced technology. Once freed of the boundaries of a single world, every option would be open. In one direction, multitudes of habitable planets would be assembled around convenient suns. The end evolution here is the Dyson Sphere, a spherical structure totally encompassing a star. The entire interior surface would be available for the biosphere. It is a concept both wildly imaginative and quaintly limited, as I shall explain.

One consequence of this kind of engineering would be the requirement to maintain very long-term narrow thermal and other radiation limits on these planets or other structures. Olaf Stapledon in the 1930s imagined habitable worlds being gradually shifted in their solar orbits to account for stellar evolution. Alternatively, the stars themselves would be engineered to promote stability and inhibit dangerous radiation outbursts. At least by this point, such engineering activities could become apparent across galactic distances. Whether these manipulations would be recognizable as artificial is a serious question for astrophysicists! And this is the key issue for SETI.

The 'signature' of ET terraforming activities and stellar engineering activities is a worthy topic for speculation. From one direction, currently conceivable activities for future terrestrial civilizations can be extrapolated until the applied energies become detectable at galactic distances, and those signals (an accidental by-product of activities with other intentions) predicted. From the other direction, observations already made can be compared against hypothesized artificial causes, such as the 'Little Green Men' suggestion for pulsars, or the only-half-whimsical suggestion that Seyfert galaxies are alien mega-Chernobyls. This last suggestion raises the serious observation that once such energies are harnessed for artificial purposes, they may not always work: recognizable ET signatures may consist of news that some civilization's greatest engineering effort has failed spectacularly. Or even more sadly, the traces may reflect energies released in an unimaginable technological conflict (what we today would call a 'war') between striving forces whose technical cleverness far outstripped their wisdom. Extrapolating from our limited sample size, that is not an inconceivable outgrowth of technological activities.

Furthermore, these signals may only represent a brief phase in the development of a technological civilization. Somewhere along these lines of technological development, ET civilizations may dive back under the cloak of long-range undetectability as they discover and harness even newer discoveries of physics. True 'artificial gravity' (not circular momentum masquerading as 'weight') may remove the need for using thousands of kilometers of inert rock to create a local acceleration field and thereby hold relatively small masses of gas and liquids. True mastery of nuclear fusion processes may remove the need for furnaces to be a million kilometers across and too hot to approach within a hundred million kilometers (a 'Dyson Sphere' thus assumes that such a civilization would have disproportionately vast abilities in one area but be unable to master stellar fusion control).

For those ET civilizations still depending on biochemical processes, habitable volumes could then be manufactured on any desired scale, from kilometers to many millions of kilometers. They need not give off long-range detectable energies of any kind. But by this stage, or soon thereafter (on a galactic timescale), the question of detecting them at galactic distances would become moot, because such evolved, advanced technologies would overcome all former limitations in original location and speed, and would no longer even need to be detectable across immense gulfs of space. They would be here. But their detectability and/ or recognizability would be entirely up to them and their unknowable motives.

Having made full circle back to Earth, we may speculate on the possible future need for deliberate human climate engineering to repair and enhance our home planet's habitability, against threats both artificial and natural. But long before this course of action was recognized, theorists had suggested in their own metaphors that Earth had already been terraformed by an extraterrestrial culture. This was the concept of 'directed panspermia', that life is not native to this planet but was imported for some alien purpose. It is possible that it has already happened here. It is possible that we ourselves will some day make it happen elsewhere. These may be the most visible marks that intelligence makes on a 'wild' galaxy, and it is for traces of these activities that, perhaps, we should search.

11

Estimates of Expansion Timescales

ERIC M. JONES

Introduction

'I see that the valleys are thick with people and even the uplands are becoming crowded. I have selected a star and beneath that star there is a land that will provide us with a peaceful home.'

Ru, Traditional Founder of Aitutaki in the Cook Islands (Buck, 1938)

An important corollary of the question 'Where are they?' is the question 'Could they have gotten here yet?' If we imagine a spacefaring civilization arisen a billion years ago and a thousand parsecs from Earth, what are the odds that the descendants of that civilization would have established settlements in the solar system before now? The answer, I believe, is that, if such a civilization had arisen and if interstellar travel is practical at a small percentage of light speed, it is virtually certain that the solar system would have been settled by non-natives long ago. Unless we discover that interstellar travel is impractical, I conclude that we are probably alone in the Galaxy.

We know nothing of any extraterrestrial civilization. If we assume that some have existed, it is also reasonable to assume that at least some would be as inquisitive and as eager for adventure as humanity (Hart, 1975; Jones, 1985). It would take but one such species to fill the Galaxy.

Humanity has a history of expansion into available areas on Earth. If we examine our past, we can estimate how long it might be before humanity would expand throughout the Galaxy. If that time is greater than a few billion years, we would conclude that the question 'Where are they?' is not very meaningful. But if the time is less than a billion years or so, the apparent absence of extraterrestrials from the solar system is significant.

I estimate that if we develop the technical means for practical interstellar travel, humanity and its descendants will fill the Galaxy in a relatively brief time – 300 million years at most, the most likely value being 60 million years. This estimate is based on assumptions about local rates of population growth, the rate at which emigrants leave one place for another, and the choice of a mathematical model. The choices are debatable. Newman & Sagan (1981) have argued for very low rates and for a different mathematical model. They estimate that the expansion time – the time to fill the Galaxy – exceeds 10^9 years. Let us examine the expansion/settlement process and the choice of rates. Afterwards, we shall briefly discuss the mathematical models and, finally, discuss estimates for the expansion time.

Human Expansion

Our ancestors seem to have begun in East Africa 1–2 million years ago. From earliest times humans have lived in small gatherer/hunter bands which diffused throughout the Old World (Finney & Jones, 1983). The dispersal across the Old World was undoubtedly unplanned. A band might chance to move a little further from East Africa because of movement of game, local climate changes, population pressures, warfare, disease or natural disaster. The decision to move was made within the band, but the net result of millions of such decisions was that our ancestors, members of a clever and adaptable species, inhabited all parts of the Earth that were physically accessible to them before the rise of agriculture some 10 000 years ago (Davis, 1974). Although humanity spread across the face of the Earth, the population increased only slowly. The rate of increase during humanity's gatherer/hunter existence has been estimated to have been only 0.0015% per year (Coale, 1974). The available technologies could support only a very modest growth rate.

Within the past 10 000 years, the idea of agriculture arose independently at several places. With agriculture came an explosion of technological development and social institutions and a gradual end to the gatherer/hunter existence. Agriculture produced food in abundance and the new social institution provided for its distribution. The population grew and concentrated in villages, towns and cities. Between 8000 B.C. and A.D. 1 the population growth rate may have been 0.4% per year (Coale, 1974).

Although humanity had largely given up the gatherer/hunter existence, there were still reasons to move. Curiosity, food shortages, overcrowding, war, religion, politics and countless other factors motivated countless emigrations. Some of the post-agricultural settlement ventures have been conducted by political institutions – notably the European colonies in North America. These were true colonies rather than settlements, in that the sponsoring institution maintained political control.

There have been two factors which have tended to assure that human settlement has largely remained a process driven by the decisions of individuals and/or small groups. First, even when oceangoing vessels were needed for transportation, they have never been so expensive that large institutions could control the means of emigration. Second, transportation and communication were so slow that control of emigration or immigration could not be maintained for long. Human institutions seem not to be able to maintain control from a distance. Local institutions evolve to meet local needs. Colonies become independent communities after they achieve self-sufficiency.

We have spread across the face of Earth and planted permanent, self-sufficient communities on all but the most inhospitable lands. Of places on Earth, only the ocean floor remains to be settled. We have no other place left to go but into space.

Settlement of the Solar System

In the near term, we shall colonize rather than settle space. Proposed large orbital habitats (O'Neill, 1974) and their support technologies could be established only at enormous expense. Only the largest of human institutions could conceivably pay the estimated 10^{11} cost of establishing the first orbital community of 10^4 colonists. Such an enormous enterprise may never be attempted. However, unless we abandon space ventures entirely, economic, scientific and political justification should exist for establishing permanently manned facilities in space and, possibly, smaller installations on the surfaces of the rocky planets and moons.

The enormous cost of lifting mass from the surface of Earth and the other large bodies in the solar system will probably guarantee that the space-living population will grow slowly *and* that efforts will be made to make the orbital facilities self-sufficient in order to reduce the costs of resupply.

Gradually, the space-living population and the technology base in space will increase. Habitats built from asteroidal or lunar materials will house small communities. Clusters of communities may occur near raw materials (the Earth–Moon system, Mars, Jupiter, the asteroidal zone), but humans will be scattered throughout the solar system, living in habitats connected by vast communications and data networks and separated by small energy differentials.

Interstellar Settlement

Once a sufficient base of population and technology is established in near-Earth orbit, the settlement of the solar system should proceed fairly steadily. Solar system distances are short enough to ensure that no journey would take more than a few years; the energy expenditures for travel would be modest. And, as in the pre-agricultural expansion of humanity across the face of the Earth, the crucial resources – sunlight, carbon, water, etc. – are accessible throughout the solar system, particularly among the asteroids and in the jovian and saturnian systems.

The next step outward, the step into interstellar space, introduces a new factor into the expansion/settlement process: extreme separation of potential settlement sites. Unless we assume that human communities can flourish in interstellar space (Dyson, 1979; Jones & Finney, 1983), the settlement sites will be orbits close enough to stars for starlight to be used as the principal energy source and for a dependable supply of raw materials in the form of asteroids, comet-nuclei, moons and gas giant atmospheres to be near at hand. If we assume that settlements will be established in orbit about single, late-type stars and that the interstellar vessels move at about one-tenth of the speed of light, the settlements will be separated, on average, by 4 parsecs and journeys of 125 years (Jones, 1981).

In no previous human expansion has the average journey been life-long. Even though the interstellar vessels may house large communities of emigrants (perhaps 10^4 or 10^5 per vessel) and resemble the interplanetary habitats in which the emigrants and their ancestors had lived for centuries, millennia or longer, and even though communications could be maintained with the home system and with the destination, the emigrants would be effectively isolated from the rest of humanity by distance and by communications time lags of up to 12 years! It is difficult to imagine the maintenance of interstellar colonies. Even if the journeys were made in suspended animation, a 125 year time gap could be a serious problem for colonial administrators. It seems evident that any human settlements established outside the solar system will, of necessity, be independent, self-sufficient communities.

An analogous human migration was recorded in part in the oral history of the Polynesian peoples (Buck, 1938). In seemingly extraordinary feats of navigation, the widely separated islands of the Pacific were settled, perhaps in several episodes, by agricultural/fisher peoples of Asiatic origin. Although there is considerable debate on many details of the settlement of Polynesia, it appears that the Pacific was settled long after the rest of the world. Settlement of Easter Island, 1100 miles from the nearest land (Pitcairn), occurred only a few hundred years before the arrival of Europeans.

The Polynesians were superb navigators. They used the stars, winds and currents, and observations of migratory birds and clouds, as they crossed hundreds of miles of open ocean between island groups. The initial discovery of an island may have involved some luck (good or ill) but, once found and reported, the islands were not lost.

The great Pacific distances required that the emigrants carry with them supplies for voyages of up to three or four weeks as well as everything they would need when they arrived at their destination. Along with themselves and their culture, they brought food plants (notably coconut, bananas and breadfruit) and domestic animals (dogs, fowl and pigs). Contact was maintained over long periods with other parts of Polynesia, but cultural and language divergence was certainly evident by the time the Europeans arrived (Buck, 1938). The Polynesian peoples and culture spread throughout the Pacific because the necessary technology was available. In some sense it was the technology that spread. Similarly, in the interstellar case, cultural and even genetic drift can be expected. If human descendants fill the Galaxy, they may well be members of diverse species and cultures but their biologic and technologic heritage will have originated in the solar system.

Mathematical Models

The dispersal of a tracer gas in a background gas is described by the diffusion equation. Each molecule of tracer moves an average distance, called the mean free path, between collisions with the background gas. The collisions are isotropic: there is no preferred scattering angle. A requisite for application of the diffusion equation is that the mean free path be short compared with any gradient length scales; the density of tracer particles cannot change by large factors over a mean free path.

There are clearly physical situations to which the diffusion equation does not apply. An obvious example is a chemical explosion in which the explosive products move outward at high speed, forming a shock wave. At the shock front the gradient length scale is comparable with the mean free path, and the approximations which lead to the diffusion equation break down.

The dispersal of plant and animal species has been successfully treated with the diffusion equation (see, for example, Newman, 1983). The modifications necessary to extend the diffusion model to living creatures are the addition of a source term (the population increases) and interpretation of the mean free path as the average distance between an individual's place of birth and the birthplace of offspring. As long as the gradient length scales remain relatively large, the modified diffusion equation is an entirely adequate description of the dispersal of life.

Newman & Sagan (1981) have attempted to apply the diffusion equation to interstellar migrations. Their equation may be written as

$$\frac{\partial P}{\partial t} = \alpha P \left(1 - P/P_s\right) + \gamma \Delta^2 \frac{\partial}{\partial x} \left(\frac{P}{P_s} \frac{\partial P}{\partial x}\right) \tag{11.1}$$

where P = the population of a settlement; P_s = the carrying capacity of a settlement; t = time; x = spatial coordinate; α = local population growth rate; γ = emigration rate, and Δ = mean separation of settlements.

The solution to eq. (11.1) is

$$P/P_s = 1 - \exp\left(\frac{x - vt}{L}\right) \tag{11.2}$$

where

$$L = \Delta \sqrt{\frac{2\gamma}{\alpha}} = \text{gradient length scale}$$

and

$$v = \sqrt{\frac{\alpha\gamma}{2}} = \text{wave speed}$$

For problems of interest the local growth rate (α) greatly exceeds the emigration rate (γ), so that $L << \Delta$. The clear implication is that the modified diffusion equation cannot apply to interstellar migrations.

In some ways interstellar migrations may resemble explosions. With relatively slow emigration rates the populations at neighboring settlements can differ by large factors. Spatial population gradients can be steep at the frontier and important to calculations of an accurate solution.

A more appropriate solution method is a discrete, Monte Carlo simulation (Jones, 1976, 1981). Briefly, we scatter settlement sites in a test volume. We assume that local population growth is given by

$$\left[\frac{\partial P}{\partial t}\right]_L = \alpha P \left(1 - P/P_s\right) \tag{11.3}$$

and that emigration is

$$\left[\frac{\partial P}{\partial t}\right]_E = \begin{cases} -\gamma P \left(1 - P/P_s\right) & P > P_L \\ 0 & P < P_L \end{cases} \tag{11.4}$$

where the parameter P_L is a threshold population such that emigration occurs when $P > P_L$, while immigration from nearby settlements occurs only when $P < P_L$. Numerical experiments demonstrate that the precise treatment used to describe emigration and immigration is relatively unimportant to the solution. More details are given by Jones (1976, 1981).

Calculational Results

Monte Carlo solutions have been obtained under these assumptions:

1. The density of settlements is 0.0015 pc^{-3}. This corresponds to a mean separation of sites of 2.2 pc.
2. The ship speed (v_s) is 0.1 c = 0.03 pc yr^{-1}.
3. The maximum voyage is 22 pc.
4. Emigrants are sent alternately to the two nearest open settlement sites ($P < P_L$).

The solutions are not particularly sensitive to the choice of a ship range unless the range is comparable with the separation of sites. If this is the case, settlement is likely not to occur; the home system and/or the first few settlements are isolated from other potential sites. The settlement wave dies off.

The solutions are presented in fig. 11.1, as calculated wave speed as a function of the coefficients α and γ, both given in parts per year. The wave speed is given in parsec per year. Newman (private communication) has shown that the Monte Carlo results are well approximated by

$$v = \overline{\delta r} / \left[(\overline{\delta x}/v_s) + (1/\alpha) \ln(2\alpha/\gamma) \right] \tag{11.5}$$

where $\overline{\delta r}$ is the average radial distance traveled, and $\overline{\delta x}/v$ is the average travel time. Usually we can assume $\overline{\delta r} = 0.7\Delta$ and neglect the travel time and use

$$v = 0.7 \ \alpha\Delta/\ln(2\alpha/\gamma). \tag{11.6}$$

Discussion

It is evident from fig. 11.1 that as long as $\gamma \ll \alpha$, the wave speed is dominated by the local growth rate (α). The evidence suggests that since the rise of agriculture the local growth rate has exceeded 10^{-3} yr^{-1} (Coale, 1974). It seems likely that the very high rates experienced since the industrial/scientific revolution (1–2% per year) are a temporary departure from normality due to the dramatic decrease in the death rate in the past 200 years. We speculate that a more appropriate rate would be $\alpha = 10^{-3}$ yr^{-1}.

If $\alpha \geq 10^{-3}$ yr^{-1}, it is evident that the wave speed will exceed 10^{-4} pc yr^{-1} unless the emigration rate is less than 10^{-8}. The choice $\alpha \geq 10^{-3}$ yr^{-1} gains some credence from the consideration that if one invokes small α to explain the absence

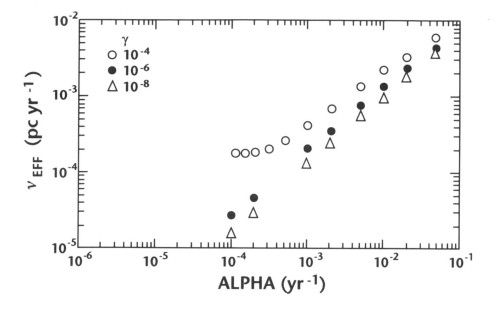

Figure 11.1 The wave speed calculated for human settlement of the Galaxy is plotted as a function of the local population growth rate (alpha) for three values of the emigration rate (gamma). Newman (private communication) has shown that the wavespeed is well approximated by $v = 0.7\alpha\Delta/\ln(2\alpha/\gamma)$, where Δ is the mean separation of settlement sites. For these calculations we assumed $\Delta = 2.2$ parsecs. The history of human migrations and population growth suggests that reasonable values of the parameters are $\alpha = 10^{-3}$ and $\gamma = 10^{-4}$. The calculated wavespeed is 5×10^{-4} parsecs per year and the implied time needed for humanity to settle the Galaxy is 60 million years.

of extraterrestrials from the solar system, then small growth rates must be invoked for *all* technological civilizations.

From data given by Potter (1965), the estimated emigration rate from Europe to North America during the eighteenth century was $3 \times 10^{-4}\,\mathrm{yr}^{-1}$. From data cited by Davis (1974), emigration in the period 1840 to 1930 often exceeded that value. The great Irish emigration of that period saw rates as large as $0.01\,\mathrm{yr}^{-1}$! About one hundred Irish arrived at American ports each day for years on end. It seems likely that $\gamma = 10^{-8}\,\mathrm{yr}^{-1}$, as suggested by Newman and Sagan (1981), is a gross underestimate of human emigration rates.

Even if γ is as small as $10^{-8}\,\mathrm{yr}^{-1}$, the wave speed will exceed $10^{-4}\,\mathrm{pc}\,\mathrm{yr}^{-1}$. Since the Galaxy is a thin disk about 3×10^4 pc in diameter, we estimate that humanity could fill the Galaxy in no more than 300 million years. Using the more likely values of $\alpha = 10^{-3}\,\mathrm{yr}^{-1}$ and $\gamma = 10^{-4}\,\mathrm{yr}^{-1}$, we get a wave speed of $5 \times 10^{-4}\,\mathrm{pc}\,\mathrm{yr}^{-1}$ and a migration timescale of 60 million years. Because both of these times are short compared with the 4.6×10^9 year age of the solar system, the apparent absence of non-native civilizations from the solar system is significant. Only if $\alpha < 10^{-4}$ do the timescales become comparable.

Caveats

Are we alone? The settlement calculations suggest that we are. However, the calculations are based on extrapolations from our past and speculations about our future. They must be viewed with suspicion.

There are several assumptions we have made which could be drastically wrong. Foremost among these is the assumption that interstellar travel is practical. Interstellar settlement will not occur on any significant scale if the voyages are very expensive. I suspect that if the labor costs exceed a few man-years per emigrant, the emigration rate will be very low, if not zero. The cost will not include the technological base. That will have to exist for its own sake. Interstellar transportation systems will have to be an outgrowth of ordinary interplanetary systems – interstellar 747s rather than interstellar Saturn Vs. I suspect that the requirement that interstellar travel not be very expensive is not a serious prerequisite. Interstellar transportation systems may seem expensive from our perspective, but, then, so would a 747 to the Wright brothers.

One further caveat needs to be mentioned. If humanity can live in interstellar space, using stars and planetary systems merely as way-stations (Dyson, 1979), our estimates of the settlement times may be meaningless. The diffusion equation will apply, Δ will be very small, and, most importantly, the absence of obvious signs of settlements in the solar system will be insignificant. Ben Finney and I have considered the question of an interstellar migration based on the settlement of comets sprinkled throughout interstellar space (Jones & Finney, 1983; Finney & Jones, 1984) and estimate that the 'filling time' in this case is the galactic radius, 10 kpc, divided by the dispersion velocity of the comet population, 30 km/s. The filling time is then 300 million years, a time still very short compared with the age of the Galaxy.

If I were forced to bet on the question, I would bet that we are alone. But we won't have a definitive answer until we have explored a large portion of the Galaxy and found no one home.

Acknowledgements

This work has greatly benefited from several lively discussions and exchanges of letters and manuscripts with W. I. Newman. My colleagues at Los Alamos have also contributed thoughts and hours of patient listening: R. W. Whitaker, M. T. Sandford II, B. W. Smith and H. G. Hughes. I gratefully acknowledge the continued interest and support of R. R. Brownlee.

The assistance of the staff of the National Library, Dublin, Ireland, is gratefully acknowledged. In search of an ancestor, I was directed to copies of nineteenth century passenger lists. An examination of the lists of people arriving at New York, Baltimore and other American ports on a single day in August 1859 gave a sense of the magnitude of the Irish emigration. These lists are also available at the National Archives, Washington, DC.

I note that Peter H. Buck, whose book *Vikings of the Pacific* is an interesting introduction to the Polynesian migration, was Maori/Irish – hence, the product of two great migrations.

This work has been supported by the Los Alamos Scientific Laboratory, which is operated by the University of California under contract to the US Department of Energy.

References

BUCK, PETER H. (1938). *Vikings of the Sunrise*. New York: Lippincott. Reissued as *Vikings of the Pacific*, University of Chicago Press, 1955.

COALE, ANSLEY J. (1974). The history of the human population. *Sci. Am.*, **231**, 40–51.

DAVIS, KINGSLEY (1974). The migrations of human populations. *Sci. Am.*, **231**, 92–105.

DYSON, FREEMAN J (1979). *Disturbing the Universe*. New York: Harper and Row.

FINNEY, B. R. & JONES, E. M. (1983). From Africa to the stars: the evolution of the exploring animal. In *Space Manufacturing 1983*, ed. J. D. Burke & A. S. Whitt, pp. 85–104. San Diego: Univelt.

FINNEY, B. R. & JONES E. M. (1984). *Interstellar Migration and the Human Experience*, Berkeley: University of California Press.

HART, MICHAEL H. (1975). An explanation for the absence of extra-terrestrials on Earth. *Quart. J. Royal Astron. Soc.*, **16**, 128–35.

JONES, ERIC M. (1976). Colonization of the galaxy. *Icarus*, **28**, 421–2.

JONES, ERIC M. (1981). Discrete calculations of interstellar migration and settlement. *Icarus*, **46**, 328–36.

JONES, E. M. (1985). Where is everybody? *Physics Today*, August 1985, 11–13.

JONES, E. M. & FINNEY, B. R. (1983), Interstellar nomads. In *Space Manufacturing 1983*, ed. J. D. Burke & A. S. Whitt. pp. 357–74. San Diego: Univelt.

NEWMAN, W. I. (1983), The long-term behaviour of the solution of a non-linear diffusion problem in population genetics and combustion. *J.Theor. Biol.*, **104**, 473–84.

NEWMAN, W. I., & SAGAN, C. (1981). Galactic civilizations: population dynamics and interstellar diffusion. *Icarus*, **46**, 293–327.

O'NEILL, G. K. (1974). The colonization of space. *Physics Today*, **27**, 32–40.

POTTER, J. (1965). The growth of population in America, 1700–1860. In *Population in History*, ed. D. V. Glass and D. E. C. Eversley, pp. 631–79. New York: Edward Arnold.

12

A Search for Tritium Sources in Our Solar System May Reveal the Presence of Space Probes from Other Stellar Systems

MICHAEL D. PAPAGIANNIS

Introduction

The possibility that life, primitive or advanced, might exist in other parts of the universe has occupied the thoughts of scientists and laymen for thousands of years. One of the earliest was the statement by the ancient Greek philosopher Metrodorus of Chios around 400 B.C., who wrote in his book *On Nature* that: 'It is unnatural in a large field to have only one shaft of wheat, and in the infinite Universe only one living world.'

In A.D. 1690 the famous Dutch physicist Christian Huygens wrote in his book *Cosmotheoros* that: 'Barren planets, deprived of living creatures that speak most eloquently of their Divine Architect, are unreasonable, wasteful and uncharacteristic of God, who has a purpose for everything.'

In the nineteenth century, several proposals were made by different distinguished scientists. The most famous was mathematician Carl Friedrich Gauss, who proposed to establish contacts with advanced civilizations on other planets of our solar system, by planting a rectangular triangle with wheat in Siberia, with squares of pine trees at its three sides, to show that the Earth has intelligent beings that know the Pythagorean Theorem. None of these proposals, however, was implemented.

The modern era of the Search for Extra-Terrestrial Intelligence (SETI) started in 1959 with a paper to *Nature* by Cocconi and Morrison, which was followed soon after in the spring of 1960 by the first radio search by Frank Drake (Project OZMA), using the then new 85 foot radio telescope at the National Radio Astronomy Observatory in West Virginia.

103

Since this first effort, the number of radio searches around the Earth have increased rapidly, including the establishment of several SETI-dedicated facilities such as the Ohio State Radio Observatory that has been operated by Professors John Kraus and Bob Dixon since the early 1970s; the Harvard University radio telescope operated by Professor Paul Horowitz since the early 1980s; and a radio facility in Argentina under the direction of Dr Raul Colomb since the early 1990s. The last two facilities are supported by the Planetary Society, and the Harvard facility is supported also by NASA. In the nearly 33 years that have passed since the first radio search by Professor Frank Drake, we have accumulated close to 33 years of observing time, but unfortunately without any positive results.

An important development in this modern period was the establishment by the International Astronomical Union (IAU), at its 1982 General Assembly, of a new Commission to search for life and intelligence in the universe. Professor Papagiannis became its first President for the period 1982–1985, with Professor Drake and Dr Kardashev as its two Vice-presidents. Professor Papagiannis introduced also the name 'Bioastronomy' for the new IAU Commission, which is now called IAU Commission 51 – BIOASTRONOMY. He also organized its first IAU Symposium (No. 112) at Boston University and edited its proceedings. He also started publishing the bulletin *Bioastronomy News*, which he edited for 10 years with the help of the Planetary Society (Papagiannis, 1984–93). In the fall of 1993 Dr Guillermo A. Lemarchand took over as editor.

What Makes a Planet Habitable?

The availability of water, H_2O, is the most important factor that makes a planet habitable, because water is a very effective polar molecule and, hence, an excellent solvent and facilitator for the complex chemistry of life. It is also made of the two most common chemically active elements, hydrogen and oxygen, and, hence, it must be quite plentiful in the universe.

The presence of water presupposes a planet with a significant mass and a substantial atmosphere. It requires also a planet with a reasonable rate of rotation to avoid overheating on the side of the central star, and freezing on the opposite side. It implies also that the planet is at a moderate distance from the central star, where water can exist in its liquid form, a range called the 'Ecosphere' or the 'Habitable Zone'. This is very important because the complex organic compounds needed by life decompose at high temperatures (since the speed of all chemical reactions doubles as the temperature is increased by 10 K), while at low temperatures all chemical reactions slow down, thus stagnating life on frozen planets.

Organic Compounds

Life is a very complex phenomenon and therefore it requires very large and complex molecules with hundreds or thousands of atoms. Such complex molecules can be constructed only around a skeleton of carbon (C) or silicon (Si) atoms, both of which have four valences, and which, as a result, can build highly complex molecules ('organic compounds' with a spine of carbon atoms, or 'silicones' with a spine of silicon atoms). However, when silicon combines with oxygen, they form SiO_2 (sand), which is not soluble in water and, as a result, stops all chemical activities. Carbon, on the other hand, when it combines with oxygen, forms CO and CO_2, both of which are gases that dissolve easily in water – especially the CO_2 which continues an active chemistry in water (soda drinks, etc.).

A characteristic advantage of carbon over silicon is the fact that, although C is 10 times more common than Si in the universe, in the crust of the Earth Si is 600 times more common than C. This is because at the temperatures of the Earth's crust the simple compounds of carbon (CO, CO_2, CH_4, etc.) are mostly gases, while the basic compound of silicon (SiO_2) is solid (sand). As a result, much more SiO_2 than CO_2 was incorporated into the solid Earth.

It is interesting to note that the chemical composition of life on Earth is very much in line with the abundances of the chemical elements in the universe. The six most common chemical elements in the universe are H, He, O, C, Ne and N. Two of them (He and Ne) are noble gases and, hence, chemically inert, while the other four (H, C, N, O), which are chemically active, account for 98% of the biomass of the Earth. It follows, therefore, that the chemical choices that life has made on Earth were very much in line with the abundances of the chemical elements in the universe, which implies that Nature has been favorably predisposed to the emergence of life on suitable planets.

Extraterrestrial Life

Life is likely to originate on planets that have moderate temperatures (~300 K), between the very hot temperatures of the stars (3 000 K–30 000 K), and the very cold temperature (3 K) of the cosmic background radiation, which is the result of the continuous expansion of the universe, which has been cooling off since its very beginning.

Life on planets with moderate temperatures (~300 K) is able to receive high-quality energy from its hot central star ($\sim6\,000$ K), and use it to grow, to multiply and to evolve. The presence of the cold (3 K) cosmic background radiation provides life with a convenient dumping ground, where it discards its low-quality infrared radiation.

A Search for Sources of Tritium in Our Solar System May Reveal the Presence of Extraterrestrial Probes

On Columbus Day, 12 October 1992, NASA initiated a major project to search for extraterrestrial intelligence in other stellar systems, up to a distance of 100 light-years from the Earth. This ambitious project was planned to continue for 10 years or more, and the hope was to study nearly 1000 Sun-like stars, some of which may be transmitting radio signals to Earth, after having detected an oxygen atmosphere on our planet. Unfortunately, funding of this ambitious project may be reduced, or even eliminated.

NASA's project, however, presumes that these advanced stellar civilizations would have already detected the emergence of a technological civilization on Earth and may be trying to establish contacts with us. We must not forget, however, that our technological civilization is only 100 years old, and in many ways may really be only 50 years old.

It is more likely that a more advanced technological civilization would know for thousands, or even millions, of years that in our solar system there is a planet with an oxygen atmosphere and, hence, with advanced life, and they may have been sending us radio signals for a long time. Their technology, moreover, is likely to be far more advanced than ours, and therefore they may have also sent to our solar system 'unmanned' space probes to investigate our Earth and report back.

These probes, traveling at speeds close to 1% of the speed of light, using nuclear fusion of hydrogen into helium for acceleration and deceleration, would be placed in orbit around our Sun and possibly around the Earth. Since these automated interstellar space probes are likely to be fusing hydrogen into helium as their energy source, as a by-product they will produce tritium, which emits at a radiofrequency of 1516.701 MHz.

Valdez & Freitas (1986) conducted in the mid-1980s a radio search for tritium in other solar systems – unfortunately, without any positive results. The absence of tritium radio signals, however, may be explained by the colossal distances that separate the Earth from other solar systems. It would be important, therefore, to conduct a search for tritium radio sources in our own solar system, to explore the possibility that space probes from other stellar systems may be in orbit around the Earth, or around the Sun, producing tritium as a by-product of the nuclear fusion of hydrogen into helium.

Since tritium has a very short half-life of only 12.5 years, if it were detected in our solar system it would imply the active presence of one or more alien space probes. This would be the easiest way to detect advanced alien technology. A search for tritium signals in our solar system, therefore, is an important objective, which may reveal the presence of extraterrestrial probes.

References

COCCONI, G. & MORRISON, P. (1959). Searching for interstellar communications. *Nature* **184**.

PAPAGIANNIS, M. (1984–93). *Bioastronomy News*. The Planetary Society.

VALDEZ, F., & FREITAS, R. A., JR. (1986). A search for tritium hyperfine line from nearby stars. *Icarus*, **65**, 152–7.

13
Primordial Organic Cosmochemistry

CYRIL PONNAMPERUMA AND
RAFAEL NAVARRO-GONZÁLEZ

Introduction

'If we could conceive, in some warm little pond, with all sorts of ammonium and phospheric salts – light, heat, electricity etc. present, that a proteine compound was chemically formed ready to undergo still more complex changes . . .'

(Charles Darwin to his friend Hooker, 1871) (Anon., 1961)

Here in a nutshell is the entire concept of chemical evolution. What the experimentalist does is to try to recreate Darwin's warm little pond and to see whether those reactions that preceded the emergence of life can be retraced in the laboratory. Such ideas lay fallow for a long period of time until the Russian biochemist Alexander Oparin, in a dissertation published in Russia in 1924, contended that there was no fundamental difference between a living organism and lifeless matter and that the complex combinations, manifestations and properties so characteristic of life must have arisen in the process of the evolution of matter (Oparin, 1924). In 1928, Haldane had similar ideas. He described the formation of a primordial broth by the action of ultraviolet light on the Earth's primitive atmosphere (Haldane, 1929). The Oparin–Haldane hypothesis is the basis of the scientific study of the origin of life.

Primitive Earth's Atmosphere

The composition of the primitive atmosphere is of paramount importance for the synthesis of organic material. The primary Earth's atmosphere was probably formed from the gravitational capture of gases from the solar nebula (Rasool, 1972); however, it was rapidly lost during the early evolution of the Sun. The secondary Earth's atmosphere originated from the outgassing of the Earth's

Table 13.1. *Energy sources available for organic synthesis on the Earth[a]*

Energy source	Flux[b] ($J\ cm^{-2}\ yr^{-1}$)	
	Present	4×10^9 yr ago
Solar radiation		
< 2500 Å	2387	> 2387
< 2000 Å	356	> 356
< 1500 Å	14.7	> 14.7
Ionizing radiation		
^{40}K, ^{232}Th, ^{235}U, ^{238}U and ^{244}Pu	3.35–64.85	11.72–196.65
Solar wind	0.84	?
Cosmic rays	0.006	0.006
Shock waves from meteorite impact	0.4	> 0.4
Heat from volcanos	0.5	> 0.5
Electric discharges:		
Lightning	0.2	0.2
Corona	0.1	0.1

[a]The flux is projected per square centimeter surface area of the Earth.
[b]Ponnamperuma *et al.* (1992).

interior (Holland, 1962) and from late accretion of cometary and meteoritic material. This secondary atmosphere was anoxic in nature; however, there is no general agreement on the actual chemical composition of the secondary atmosphere. There are basically two different points of view: a mildly reduced atmosphere (methane, nitrogen and water) or a neutral atmosphere (carbon dioxide, nitrogen and water). The experimental evidence indicates that a mildly reduced atmosphere is more efficient in synthesizing a larger variety and quantity of organic compounds than is a neutral atmosphere.

Energy Sources

Another important factor for the process of chemical evolution is the type of energy sources available to initiate the synthesis of organic compounds (Miller & Urey, 1959). Table 13.1 summarizes the energies available on the Earth. The principal source of energy on our planet, now and in the remote past, is that from the Sun. The most important region in the spectrum of solar radiation is ultraviolet light because of its ability to dissociate the gas components of the primitive Earth. Ionizing radiations from cosmic rays and from the decay of radioactive elements in the lithosphere and hydrosphere were probably another energy source on the primitive Earth. Heat emitted from lava was a significant but not a major source of energy. Shock waves from the impact of comets and meteorites on the Earth's atmosphere as well as those generated in lightning bolts have also been proposed as sources of heat to initiate chemical reactions.

Another significant source of energy would be electric discharges resulting from coronal discharges and lightning in the atmosphere.

Simulation Experiments

The earliest published experiment expressly designed to demonstrate the formation of organic compounds from a hypothetical primitive atmosphere was that by Groth & Suess in 1938. They exposed a CO_2–H_2O gas mixture to ultraviolet light at 147 nm, and identified formaldehyde and glyoxal as products. These results were interpreted as 'giving an explanation for the formation of certain carbon compounds that were probably prerequisite for the appearance of life' (Groth & Suess, 1938). Later, Calvin and coworkers simulated the atmosphere–hydrosphere interface, by irradiating with α-rays an aqueous solution containing ferrous ions in equilibrium with a CO_2–H_2 gas mixture (Garrison et al., 1951). Formaldehyde, formic acid and succinic acid were identified among the products. Shortly after, Miller (1953, 1957a, b) reported the synthesis of amino acids such as glycine, alanine and β-alanine, among others, from the electric discharge of a CH_4–NH_3–H_2–H_2O gas mixture. Since that time, innumerable other investigators have attempted to retrace the path from a primitive atmosphere to nucleic acid-like molecules, polypeptides and fatty acids. In the majority of these studies, two precursors have been found to play a key role in the syntheses of biomolecules: formaldehyde and hydrogen cyanide.

A typical laboratory apparatus is illustrated in fig. 13.1. The upper flask represents the atmosphere, the lower flask the ocean. The side arm is hot, the condenser cold. Circulation is thus established, portraying the interaction between the atmosphere and the hydrosphere. In this particular apparatus an electrical discharge is used. At the end of a 24 h experiment, 95% of the starting methane has been converted into organic compounds. By careful analysis of the dark-brown material which has been deposited on the flask and dissolved in the water, a number of molecules of biological significance can be isolated (Ponnamperuma et al., 1969).

With different types of energy acting on a primitive atmosphere, whether of an

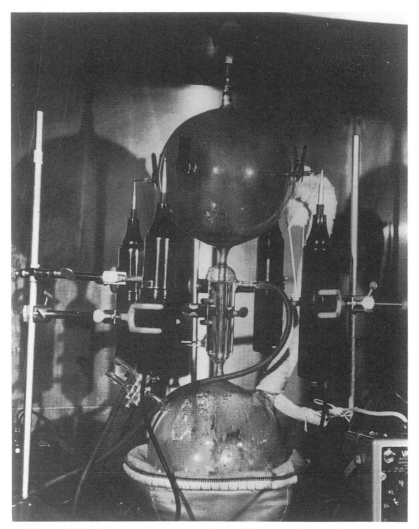

Figure 13.1 Laboratory representation of the atmosphere and the ocean.

intensely reducing or non-oxidizing nature, both hydrogen cyanide and formaldehyde appear to be formed (table 13.2).

To the chemist, prebiotic synthesis appears as a two-part problem: (1) to make
the biomonomers necessary for life; (2) to combine them under similar conditions
into the biopolymers.

Table 13.2. *Rates of productions of HCN and HCHO in the primitive atmosphere by different energy sources ($\times$ 10^9 mol yr^{-1})*[a]

Energy source	Product	Nature of the atmospheric carbon		
		CH_4	CO	CO_2
Electric discharge	HCHO	0.5–6	6–21	0.2–0.5
	HCN	500	50	5
Ultraviolet light	HCHO	10–100	5000	0.002–3000
	HCN	0	0	0

[a]Raulin (1990); Navarro-González (1992).

Synthesis of Biomonomers

We describe here the pathways for the synthesis of biomonomers using formaldehyde and hydrogen cyanide as the starting materials. A classic reaction in which both of these molecules take part is the Strecker reaction. This reaction has been suggested by Miller as a possible mechanism for the synthesis of α-amino acids in the primitive ocean (reactions 13.1–13.3) (Miller, 1957a). A variation of the Strecker synthesis is the cyanohydrin reaction which leads to α-hydroxy acids (reactions 13.1 and 13.4).

$$HCHO + HCN \longrightarrow HOCH_2CN \qquad (13.1)$$
$$HOCH_2CN + NH_3 \longrightarrow NH_2CH_2CN + H_2O \qquad (13.2)$$
$$NH_2CH_2CN + 2H_2O \longrightarrow NH_2CH_2CO_2H + NH_3 \qquad (13.3)$$
$$HOCH_2CN + 2H_2O \longrightarrow HOCH_2CO_2H + NH_3 \qquad (13.4)$$

Formaldehyde polymerizes in basic media, leading to a large variety of C_3–C_6 sugars. This process is called the formose reaction and has been frequently claimed as the main prebiotic route to ribose in the primitive ocean. However, a critical evaluation of the reaction conditions indicates that this was not a viable pathway for ribose synthesis since the reaction requires both a high concentration of HCHO and a high pH. Furthermore, a low yield of ribose is obtained, less than 1% (Shapiro, 1988). The formose reaction at lower concentrations has been reported (Gabel & Ponnamperuma, 1967; Reid & Orgel, 1967); however, a re-examination of these experiments is required with modern analytical techniques. The lack of a plausible pathway for ribose synthesis has led Schwartz and coworkers to suggest that the genetic system originated from acyclic prochiral

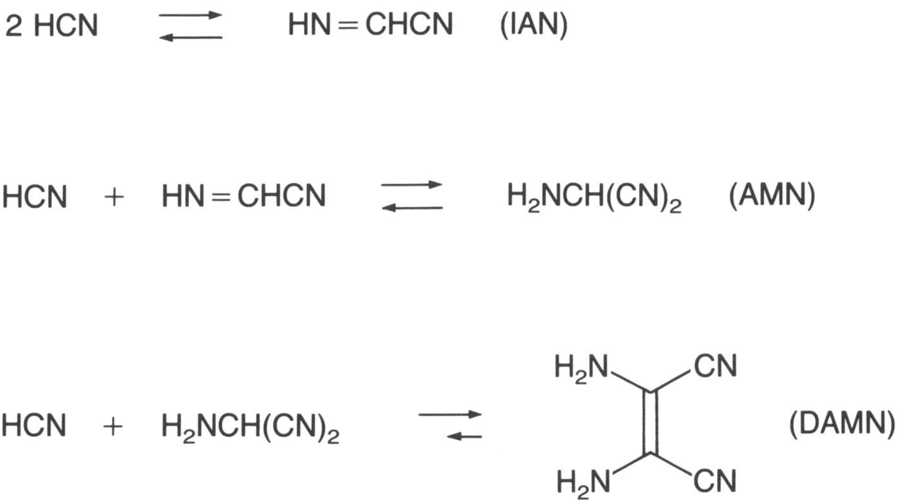

$$2 \ HCN \quad \rightleftharpoons \quad HN = CHCN \quad (IAN)$$

$$HCN \quad + \quad HN = CHCN \quad \rightleftharpoons \quad H_2NCH(CN)_2 \quad (AMN)$$

$$HCN \quad + \quad H_2NCH(CN)_2 \quad \rightleftharpoons \quad \underset{H_2N}{\overset{H_2N}{>}}C = C \underset{CN}{\overset{CN}{<}} \quad (DAMN)$$

Figure 13.2 Formation of DAMN: initial steps in the ionic polymerization of HCN (Ferris & Hagan, 1984).

nucleotide analogs that could be more easily synthesized, such as those composed of C_3–C_4 sugars (Schwartz & Orgel, 1985).

Hydrogen cyanide also polymerizes in basic media giving rise to a variety of polymers. Diaminomaleonitrile (DAMN), a tetramer of HCN, is the smallest polymer isolable (Sanchez et al., 1967). The initial steps leading to the formation of DAMN are shown in fig. 13.2. The rate of polymerization of HCN has been found to be largely enhanced in the presence of HCHO or glyconitrile (formed by reaction 13.1) (Schwartz & Goverde, 1982; Schwartz, 1983). The mechanism of this enhancement is still under investigation by Schwartz.

Another important polymer is adenine, a pentamer of HCN. This compound was first detected by Oró (1960) when a concentrated ammoniacal solution of HCN was heated for from one to several days at temperatures ranging from 27 °C to 100 °C. Ponnamperuma and coworkers (1963) synthesized adenine by irradiating with high-energy electrons a presumed primitive atmosphere composed of CH_4–NH_3–H_2O. It was suggested that adenine formed from the polymerization of HCN. Ferris et al. (1978), working with dilute HCN solutions at room temperature found that adenine could be detected only after acid hydrolysis of the solution. This observation suggests that adenine is formed as a substituted derivative or is chemically associated with higher HCN polymers. Several mechanisms have been proposed for its formation (Oró, 1961b; Kliss & Matthews, 1962; Sanchez et al., 1967; Ferris et al., 1978); however, Voet & Schwartz (1983) have recently isolated

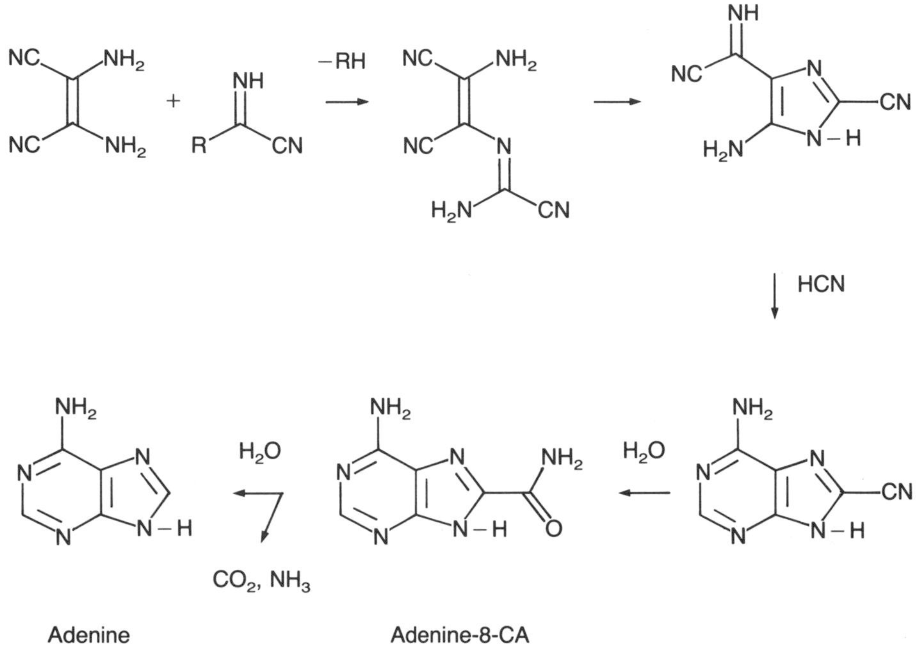

Figure 13.3　Mechanism of adenine synthesis according to Voet & Schwartz (1983).

and identified adenine-8-carboxamide in a polymerizing HCN solution. This purine derivative releases adenine upon hydrolysis. On the basis of this finding, Voet & Schwartz have proposed an original mechanism for the formation of adenine (fig. 13.3).

The structure and mechanism of formation of higher polymers is much less clear. Subsequent polymerization steps appear to involve a series of internal redox reactions in the polymer hydrolysate (Ferris & Hagan, 1984). The elucidation of the structure of HCN polymers is complicated by the heterogeneous nature of the material, and only limited progress has been made in this area. At present it is clear that higher polymers of HCN contain a more complex variety of functional units. This complexity is apparent in the diverse mixture of products that have been identified upon hydrolysis of the polymers, such as amino acids, purines, pyrimidines, hydantoins, ureas and carboxylic acids.

Synthesis of Biopolymers

The transition from the monomers to the polymers is mediated by a dehydration condensation reaction. Such a dehydration could have taken place on the primitive Earth when the mixture of organic compounds, brought to the ocean shoreline and absorbed on a clay surface, was acted upon by solar heat. This idea of polymerization on the ocean shoreline, or on the dried-up bed of a primordial lagoon, has been successfully tested in the laboratory. However, it would be more plausible if such reactions could readily take place under aqueous conditions, since the Earth is a very wet planet. Such a situation might give rise to a much more general explanation of the origin of polymers.

In attempting to simulate the dry ocean bed a nucleoside and a phosphate were mixed together and heated to about 125 °C, whereby a large number of compounds were synthesized. Among them dinucleoside monophosphates, mononucleotides and trinucleotides were found (Ponnamperuma & Mack, 1965). More recent work using calcium phosphate as a phosphorylating agent enabled us to synthesize polymers of up to 10 units.

In examining the polymerization reaction under prebiotic conditions, it is clear that a number of condensing agents could have given rise to these results (Hulshof & Ponnamperuma, 1976; Ponnamperuma, 1978). Cyanamide is the simplest of the reagents that have been used. It is formed by the irradiation of a primitive atmosphere (Schimpl & Calvin, 1965). In one of the first experiments done under aqueous conditions, glycine and leucine were irradiated with UV light in the presence of cyanamide. Among the products $(Gly)_2$, $(Gly)_3$, Leu–Gly and Gly–Leu were identified (Ponnamperuma & Peterson, 1965).

In other experiments carbodiimides have been used. The solvents here are nonaqueous in nature and may not have a direct bearing on primordial chemistry (Sheehan et al., 1967). Calvin & Steinman have examined the possible role of dicyanamide in prebiotic synthesis and have shown that polypeptides could be formed at a low pH by such a reaction (Steinman et al., 1965).

In one of our earlier investigations of the action of an electrical discharge on a primitive atmosphere, no amino acids were detected by paper chromatography: when the resulting solution was hydrolysed, 10 amino acids were identified. It could be argued that the amino acids were a result of the hydrolysis of nitriles. When enzymatic hydrolysis was used, a number of amino acids were released, suggesting the presence of the peptide bond. The peptide bond appeared to be already formed after a 24 h experiment (Flores & Ponnamperuma, 1972). Recently we have fractionated these polymers and subjected them to peptide sequencing (Su et al., 1989). These are small polymers with a maximum molecular weight of 1050. Some of these polymers have catalytic activity suggesting that primitive 'enzymes' may have emerged possibly at a very early stage of chemical evolution. An interesting conclusion from this result is that a long period of time was not necessary for chemical evolution. If the right molecules were present, and if organization took place, then the appearance of life could perhaps be 'instant in geochemical terms'.

The Role of Trace Metal Ions

Calcium, cadmium, copper, iron, manganese, molybdenum, nickel, selenium and zinc are known to be present in the active sites of a number of enzymes known as metalloenzymes. The role of these elements in catalysis is crucial in electron transfer reactions (e.g., Fe, Cu, Mo) or in acid catalyzed reactions acting as Lewis acids (e.g., Zn and Mn). Most of these elements are known to be essential to higher animals; however, Fe, Zn and Mo are essential to all living organisms (Kobayashi & Ponnamperuma, 1985). Therefore, it is logical to conclude that the need for these elements in metabolic processes arose during the process of chemical evolution and the origin of life (Navarro-González & Ponnamperuma, 1995). It may be postulated that the essentiality of only certain elements was determined by four factors:

1. Abundance. From a variety of elements with similar properties, only the most abundant has the potential to become essential.
2. Efficiency. If there are more than one element with similar abundance and properties, then the most efficient is selected.
3. Fitness. From a vast reservoir of elements, selection is based on their ability to catalyze chemical reactions of prebiotic and/or biologic importance.
4. Evolution. Metal ions probably played an important role in early chemical evolution in the form of simple complexes. The development of more complex ligands led to the improvement of catalytic functions and was inherently related to the process of the origin of life.

There are several studies on the catalytic role of trace elements in chemical evolution (for a review see Kobayashi & Ponnamperuma, 1985; Navarro-González & Ponnamperuma, 1995). In our laboratory we have studied the possible catalytic role of the cyanocomplexes of iron in the free radical polymerization of hydrogen cyanide (Navarro-González et al., 1989). Our results indicated that cyanocomplexes play no role in promoting prebiotic polymerization reactions via free radicals. There is experimental evidence indicating that the free radicals produced from HCN readily reacted by electron transfer with the cyanocomplexes. This reaction, which led to the quenching of the free radicals, produced an intermediate that, rather than undergoing addition reactions with HCN, was unstable and decomposed.

Ferrocyanide has been found to possess enzyme-like activity. Furthermore, it has been postulated as an intermediate in the evolution of iron-containing enzymes (Kamaludin *et al.*, 1986).

Role of Phosphorus

Because of its low cosmic abundance and its ready precipitation from solution as apatite, the obvious importance of phosphorus in evolution was considered an enigma (Gulick, 1955). Nevertheless, biology uses phosphorus and therefore geology will have to accommodate it. Phosphates would not have played a vital role in living systems if the phosphorus was not readily available.

If phosphorus were present in one form or another in the crust of the Earth, then, during the course of the heating generated by planetary accretion, polyphosphates would have been produced. If the oceans were formed by the outgassing of the crust, these polyphosphates would have leached into the oceans and would have been productive as condensing agents. A further argument that comes to mind from the discussion of planetary formation is a contribution from late accretion. Even after the oceans were formed, there was much scavenging in the solar system. Some of the falling debris from the primordial dust cloud, the mineral schribezite, for example, would have provided some phosphides. The water reacting with phosphides would give rise to phosphine, and to phosphate and to polyphosphates.

Role of Clay Minerals

The importance of clays in chemical evolution was first suggested by Bernal (1951). Since then, a number of experiments have been done testing Bernal's ideas (Ponnamperuma *et al.*, 1982). On the basis of these results, the role of clays in the origin of life can be described as follows:

1. Clay minerals facilitate, aid or speed the reactions in which biological monomers were formed from gaseous constituents of the primordial atmosphere (Shimoyama *et al.*, 1978). The presence of clays in the reaction medium does not change the path of the reaction but apparently promotes the reaction synthesis.
2. Clay minerals adsorbed biomonomers on their surface, providing a highly concentrated system in which monomers had specific orientations. Selective adsorption of monomers occurs only if there are large differences in the isoelectric points of the monomers; however, no distinction is made in the adsorption of biological and nonbiological molecules (Friebele *et al.*, 1980). Clays do not appear to have significant ability to sort out L and D stereoisomers (Friebele *et al.*, 1981).

3. Clays have been postulated to promote polymerization of biological monomers by concentrating them, providing containment and a surface for immobilization of the reacting molecules, and protection for reaction products in the interlayer space (Ponnamperuma et al., 1982). However, there are some reports that clays may even inhibit free radical polymerization reactions of prebiotic significance (Navarro-González et al., 1990; Negrón-Mendoza & Navarro-González, 1990). At present, there is no evidence, and, further, no theoretical reason, for the initiation of condensation reactions between biological monomers by clays.

Cairns-Smith (1982) has suggested an alternative for the origin of life. He proposes that life on Earth evolved through natural selection from inorganic crystals by a process he refers to as genetic takeover: a transformation in the biochemistry of organisms from inorganic to biologic.

Interaction of Amino Acids and Nucleotides

The next stage in the study of chemical evolution revolves around interaction between nucleic acids and proteins. Under the most primitive conditions, there must have arisen some possible relationship between polynucleotides generated under abiogenic conditions and polypeptides similarly formed. Today's genetic code is a complex apparatus that requires several steps from transcription to translation. But perhaps, during the very early conditions, there may have been some relationship between the amino acids and mononucleotides, dinucleotides or trinucleotides. Several studies have shown that there are measurable interactions between amino acids and nucleotides in solution. It was recently shown in this laboratory that amino acids have higher association constants with anticodonic nucleotides than with codonic nucleotides (Senaratne, 1986). It is strongly suggestive that the genetic code has its foundation in interactions between amino acids and nucleotides comprising their anticodonic sequences. Other factors undoubtedly must have played a part, such as some combinations of anticodon mediation with codon/anticodon.

Chirality

The biosphere is composed of asymmetric building blocks: L-amino acids (making up the proteins and enzymes) and D-sugars (making up the backbone of DNA and RNA). The questions as to how chirality arose in the biosphere and what the driving force was to push the decision in just one direction and not others are still unsolved (for recent reviews see Ponnamperuma et al., 1990; Bonner, 1991). It is clear that the origin of life is integrally related to the origin of optical activity.

Our attempts to understand both problems have been the synthetic, the analytical and the computational approaches.

In the synthetic approach, we have given particular attention to the analysis of enantiomeric distribution of amino acids formed in electric discharge experiments. In this study, it was demonstrated that equal amounts of L- and D-enantiomers are produced in the reaction medium (Shimoyama et al., 1978). The possible selective adsorption of optical isomers by clays was also examined (Friebele et al., 1980). The results suggest that apparently clays have no significant ability to sort out L- or D-amino acids. Studies on the interaction between amino acids and nucleotides have provided a possible physical and chemical basis for the origin of the genetic code (Senaratne, 1986). These results indicate that anticodonic nucleotides have higher association constants with L-homocodonic amino acids than with D-forms.

In the analytical approach, we have been interested in examining the enantiomeric distribution of extraterrestrial amino acids found in meteorites. The results show conclusively the presence of equal amounts of L- and D-amino acids in the Murchison (Kvenvolden et al., 1971), the Murray (Lawless et al., 1971), the Mighei, the Alan Hills, and the Yamato (Ponnamperuma, 1972) meteorites.

In the computational approach, we have developed a computer model to examine whether comets could protect homochiral molecules from cosmic radiation (Navarro-González et al., 1992). Bonner (1991) recently suggested that the circularly polarized light from the synchrotron radiation of neutron star remnants of supernova events may have led to molecular homochirality of biomolecules, e.g., amino acids, in interstellar grains. These homochiral molecules were probably incorporated into comets since they are believed to have formed by accretion of grain particles in the outer regions of the presolar nebula around 4.6 billion years ago. Our results suggest that if such a scenario indeed took place, cosmic rays and the internal ionizing radiation resulting from the decay of radionuclides in comet nuclei did not lead to a significant degree of racemization of the biomolecules possibly present in the interiors of comet nuclei. Therefore, comets are possible carriers of extraterrestrial sources of homochirality.

Evidence in Ancient Rocks and Sediments

The experiments that have been described so far would appear to be more credible if we had some way of finding out whether some of the primordial soup, or some of the molecules from the early prebiotic era, could be found hidden away in the Earth or on a planetary surface since the early stages of the evolution of the solar system. Several sites of Precambrian rocks and sediments are available on the present Earth, such as the Bitter Springs in Australia (1 billion years old), the Gunflint chert in Ontario, Canada (2 billion years old), the Fig Tree and the Onverwacht Formations in South Africa (3.1 to 3.5 billion years old), and the Isua rocks in Greenland (3.8 billion years old). The studies by micropaleontologists have taken us back some 3500 million years (My), where the presence of microfossils in sediments from South Africa and Western Australia have indicated the presence of life (Saxinger & Ponnamperuma, 1974; Mizutani & Ponnamperuma, 1977; Brasier, 1979). Since the Earth is $c.$ 4600 My old, we are particularly interested in sediments of the first 10^9 years of the Earth's existence, where organic matter of primordial nature may be preserved for us. Our search for these ancient sediments, especially those which may be described as 'molecular fossils', has taken us back 3800 My to the Isua supracrustal formation of western Greenland (Walters et al., 1981). Here there are sedimentary rocks which are so far the oldest known on the Earth. They contain graphitic material and, although they have been metamorphosed by various earth movements, organic molecules have been found in them. The isotopic ratio is strongly suggestive of a flourishing biota. While these observations need further confirmation, we are able to assert that life on Earth has existed for about 3800 My or that life is at least as old as the oldest known sediments (Nisbet, 1985).

Evidence beyond the Earth

Searches for prebiological matter have also been made in the lunar samples. Each of the samples from Apollo 11 to Apollo 17 has been analyzed in our laboratory. An extensive search for traces of organic molecules which may be suggestive of chemical evolution has been made. Most of this work points to about 200 ppm of carbon in the samples, but with no evidence for any molecules of organic significance (Chang et al., 1971).

Turning from lunar samples to meteorites, we have had the opportunity of examining several carbonaceous chondrites. This category of meteorites has carbonaceous material ranging from 0.5% to about 5.5%. The Murchison meteorite fell in Australia on 28 September 1969. A fragment of that sample was analyzed after its fall, and for the first time we were able to establish the presence of equal amounts of D- and L-isomers of amino acids (Kvenvolden et al., 1971). A large number of amino acids were found and among them were several which are not found in proteins. Other evidence for the extraterrestrial nature of the organic matter came from the aliphatic hydrocarbons (which are very similar to those synthesized in prebiotic experiments), the aromatic hydrocarbons (the distribution of which is very similar to a random synthesis from methane or from acetylene) and, finally, from the $\delta^{13}C$ value of the carbonaceous material which is very different from organic matter in terrestrial sediments (Kvenvolden *et al.*, 1970). This was the first conclusive proof of the extraterrestrial nature of the amino acids in a carbonaceous chondrite. This analysis has now been extended to other meteorites: the Alan Hills, the Allende, the Mighei, the Murray and the Yamato (Kvenvolden *et al.*, 1971; Lawless *et al.*, 1971; Ponnamperuma, 1972). Investigations of the extractable organic material have led to the identification of many different classes of compounds, such as hydrocarbons, alcohols, aldehydes, ketones, carboxylic acids, amines, amino acids, purines and pyrimidines (Ponnamperuma *et al.*, 1992).

The Voyager Mission to Jupiter has sent back information concerning the chemistry of the jovian atmosphere. Infrared spectra indicate the presence of various organic molecules (Hanel, 1979). Laboratory experiments in support of such possibilities have led us to the conclusion that the colors of the planet Jupiter may be a result of organic synthesis (Ponnamperuma, 1976). This process is governed by the gas phase chemistry of methane in the upper troposphere and stratosphere of giant planets (Bossard *et al.*, 1986). Laboratory experiments simulating the atmospheric conditions of these planets and/or some of their satellites have conclusively demonstrated the synthesis of complex organics such as hydrocarbons, nitriles, and sulfur- and phosphorus-containing compounds.

Further dramatic evidence to bolster the concept of chemical evolution has come to us from the interstellar molecules which radio astronomers have been discovering over the past few decades (Irvine *et al.*, 1991). Since 1968, many of the molecules that are important to chemical evolution, such as ammonia, water, hydrogen cyanide and formaldehyde, have been found. A recent count gives almost eight dozen different molecules discovered in the interstellar medium (Ponnamperuma *et al.*, 1992). There are two types of chemical processes that can synthesize molecules from atoms in interstellar clouds (Herbst, 1990): (1) gas phase reactions and (2) reactions on the surfaces of interstellar dust particles. In particular, the latter process could lead to the formation of complex organic molecules relevant for the emergence of life (Greenberg, 1989; Bonner, 1991).

Since interstellar clouds are birthplaces of stars and planetary systems, a question that immediately arises is whether or not interstellar molecules play a role in the origin of life on Earth and elsewhere. The study of comets may provide an answer to this question. Comets are considered to be frozen remnants of the solar nebula which formed at a very great distance from the Sun. As early as 1961, Oró suggested that comets could have supplied part of the initial inventory of organic matter for chemical evolution (Oró, 1961a). Presumptive evidence now exists that not only monomeric material (such as amino acids, pyrimidines, etc.) but also polymeric material constituted a part of the interstellar material transported to the early Earth via comets (Greenberg, 1989).

The laboratory experiments, the analysis of meteorites, the detection of interstellar molecules and the current data available from other planets and satellites lead us to the belief that chemical evolution is truly cosmic in nature and that primordial organic chemistry may be described as organic cosmochemistry.

References

ANON. (1961). *Charles Darwin, Life and Letters*, Vol. 3, p. 18 (*Notes and Memoirs of the Royal Society, London*, **14**, No. 1).

BERNAL, J. D. (1951). *The Physical Basis of Life*, p. 34. London: Routledge & Kegan Paul.

BONNER, W. A. (1991). *Origins of Life*, **21**, 59.

BOSSARD, A., KAMAGA, R. & RAULIN, F. (1986). *Icarus*, **67**, 305.

BRASIER, M. D. (1979). *Chem. Brit.*, **15**, 588.

CAIRNS-SMITH, A. G. (1982). *Genetic Takeover and the Mineral Origins of Life*. Cambridge: Cambridge University Press.

CHANG, S., KVENVOLDEN, K. A., LAWLESS, J. & PONNAMPERUMA, C. (1971). *Science*, **171**, 474.

FERRIS, J. P. & HAGAN, Jr. W. J. (1984). *Tetrahedron*, **40**, 1093.

FERRIS, J. P., JOSHI, P. C., EDELSON, E. H. & LAWLESS, J. G. (1978). *J. Mol. Evol.*, **11**, 293.

FLORES, J. & PONNAMPERUMA, C. (1972). *J. Mol. Evol.*, **2**, 9.

FRIEBELE, E., SHIMOYAMA, A., HARE, P. E. & PONNAMPERUMA, C. (1981). *Origins of Life*, **11**, 173.

FRIEBELE, E., SHIMOYAMA, A. & PONNAMPERUMA, C. (1980). *J. Mol. Evol.*, **16**, 269.

GABEL, N. W. & PONNAMPERUMA, C. (1967). *Nature*, **216**, 453.

GARRISON, W. M., MORRISON, D. C., HAMILTON, J. G., BENSON, A. A. & CALVIN, M. (1951). *Science*, **114**, 416.

GREENBERG, J. M. (1989). *Adv. Space Research*, **9** (6), 15.

GROTH, W. & SUESS, H. (1938). *Naturwissenschaften*, **26**, 77.

GULICK, A. (1955). *Am. Sci.*, **43**, 479.

HALDANE, J. B. S. (1929). The Rationalist Annual, **148**, 3. Republished by J. D. Bernal in *The Origin of Life*, p. 242, Weidenfeld and Nicolson, London.

HANEL, R. (1979). *Science*, **204**, 972.

HERBST, E. (1990). *Angew. Chem. Int. Ed.*, **29**, 595.

HOLLAND, H. D. (1962). In *Petrologic Studies: A Volume to Honor A. F. Buddington*, ed. A. E. Engel *et al.*, p. 447. New York: Geological Society of America.

HULSHOF, J. & PONNAMPERUMA, C. (1976). *Origins of Life*, **7**, 197.

IRVINE, W. M., OHISHI, M. & KAIFU, N. (1991). *Icarus*, **91**, 2.

KAMALUDIN, SINGH, M. & DEOPUJARI, S. W. (1986). *Origins of Life*, **17**, 59.

KLISS, R. M. & MATTHEWS, C. N. (1962). *Proc. Natl. Acad. Sci. USA*, **48**, 1130.

KOBAYASHI, K. & PONNAMPERUMA, C. (1985). *Origins of Life*, **16**, 41.

KVENVOLDEN, K. A. *et al.* (1970). *Nature*, **288**, 923.

KVENVOLDEN, K. A., LAWLESS, J. G. & PONNAMPERUMA, C. (1971). *Proc. Natl. Acad. Sci. USA*, **68**, 486.

LAWLESS, J. G., KVENVOLDEN, K. A., PETERSON, E., PONNAMPERUMA, C. & MOORE, C. (1971). *Science*, **173**, 626.

MILLER, S. L. (1953). *Science*, **117**, 528.

MILLER, S. L. (1957a). *Biochim. Biophys. Acta*, **23**, 480.

MILLER, S. L. (1957b). *Ann. NY Acad. Sci.*, **69**, 260.

MILLER, S. L. & UREY, H. C. (1959). *Science*, **130**, 245.

MIZUTANI, H. & PONNAMPERUMA, C. (1977). *Origins of Life*, **8**, 183.

NAVARRO-GONZÁLEZ, R. (1992). In *Proc. 3rd Int. Conf. Role of Formaldehyde in Biological Systems. Methylation and Demethylation Processes*, ed. E. Tyihák, p. 93. Hungarian Biochemical Society.

NAVARRO-GONZÁLEZ, R., KHANNA, R. K. & PONNAMPERUMA, C. (1992). *Origins of Life*, **21**, 359.

NAVARRO-GONZÁLEZ R., NEGRÓN-MENDOZA, A., AGUIRRE-CALDERÓN, M. E. & PONNAMPERUMA, C. (1989). *Adv. Space Res.*, **9** (6), 57.

NAVARRO-GONZÁLEZ, R. NEGRÓN-MENDOZA, A., RAMOS, S. & PONNAMPERUMA, C. (1990). *Sci. Géol., Mém.*, **85**, 55.

NAVARRO-GONZÁLEZ, R. & PONNAMPERUMA C. (1995). *Adv. Space Res.*, **15**(3), 357.

NEGRÓN-MENDOZA, A. & NAVARRO-GONZÁLEZ, R. (1990). *Origins of Life*, **20**, 377.

NISBET, E. G. (1985). *J. Mol. Evol.*, **21**, 289.

OPARIN, A. I. (1924). *Proiskhozhdenie Zhizny*, Moscow. Izd. Moskovskii Rabochii. Translated by J. D. Bernal in *The Origin of Life*, p. 199, Weidenfeld and Nicolson, London, 1967.

ORÓ, J. (1960). *Biochem. Biophys. Res. Comm.*, **2**, 407.

ORÓ, J. (1961a). *Nature*, **190**, 389.

ORÓ, J. (1961b). *Nature*, **191**, 1193.

PONNAMPERUMA, C. (1972). *Ann. NY Acad. Sci.*, **194**, 56.

PONNAMPERUMA, C. (1976). *Icarus*, **29**, 321.

PONNAMPERUMA, C. (1978). In *Origin of Life*, ed. H. Noda, p. 67. Tokyo: Center for Academic Publications, Japan Scientific Societies Press.

PONNAMPERUMA, C., HONDA, Y. & NAVARRO-GONZÁLEZ, R. (1990). In *Symmetries in Science IV*, ed. B. Gruber and J. H. Yopp, p. 193. New York: Plenum Press.

PONNAMPERUMA, C., HONDA, Y. & NAVARRO-GONZÁLEZ, R. (1992). *J. Brit. Interplanetary Soc.*, **45**, 241.

PONNAMPERUMA, C., LEMMON, R. M., MARINER, R. & CALVIN, M. (1963). *Proc. Natl. Acad. Sci. USA*, **49**, 737.

PONNAMPERUMA, C. & MACK, R. (1965). *Science*, **148**, 1221.

PONNAMPERUMA, C. & PETERSON, E. (1965). *Science*, **147**, 1572.

PONNAMPERUMA, C., SHIMOYAMA, A. & FRIEBELE, E. (1982). *Origins of Life*, **12**, 9.

PONNAMPERUMA, C., WOELLER, F., FLORES, J., ROMIEZ, M. & ALLEN, W. (1969). *Adv. Chem. Ser.*, **80**, 280.

RASOOL, S. I. (1972). In *Exobiology*, ed. C. Ponnamperuma, p. 369. Amsterdam: North-Holland.

RAULIN, F. (1990). *J. Brit. Interplanetary Soc.*, **43**, 39.

REID, C. & ORGEL, L. E. (1967). *Nature*, **216**, 455.

SANCHEZ, R. A., FERRIS, J. P. & ORGEL, L. E. (1967). *J. Mol. Biol.*, **30**, 223.

SAXINGER, C. & PONNAMPERUMA, C. (1974). *Origins of Life*, **5**, 189.

SCHIMPL, A., LEMMON, R. M. & CALVIN, M. (1965). *Science*, **147**, 149.

SCHWARTZ, A. W. (1983). *Naturwissenschaften*, **70**, 373.

SCHWARTZ, A. W. & GOVERDE, M. (1982). *J. Mol. Evol.*, **18**, 351.

SCHWARTZ, A. W. & ORGEL, L. E. (1985). *Science*, **228**, 585.

SENARATNE, S. M. (1986). *Direct Interactions between Amino Acids and Nucleotides as a Possible Explanation for the Origin of the Genetic Code*. Doctoral Dissertation, University of Maryland.

SHAPIRO, R. (1988). *Origins of Life*, **18**, 71.

SHEEHAN, J. C., GOODMAN, M. & HESS, G. P. (1967). *J. Am. Chem. Soc.*, **78**, 1367.

SHIMOYAMA, A., BLAIR, N. & PONNAMPERUMA, C. (1978). In *Origin of Life*, ed. H. Noda, p. 95. Tokyo: Center for Academic Publications, Japan Scientific Societies Press.

STEINMAN, G., LEMMON, R. M. & CALVIN, M. (1965). *Science*, **147**, 1574.

SU, Y., HONDA, Y., HARE, P. E. & PONNAMPERUMA, C. (1989). *Origins of Life*, **19**, 237.

VOET, A. B. & SCHWARTZ, A. W. (1983). *Bioorg. Chem.*, **12**, 8.

WALTERS, C., SHIMOYAMA, A. & PONNAMPERUMA, C. (1981). In *Origin of Life*, ed. Y. Wolman, p. 473. Dordrecht: D. Reidel.

14

Chance and the Origin of Life*

EDWARD ARGYLE

Introduction

It is a widely held view that life will arise spontaneously on the surface of any planet that provides a suitable physical and chemical environment. This belief is saved from tautology by the generously broad definitions of 'suitable' that abound in discussions of the origin of life. Indeed it is almost sufficient to require only that liquid water occur on the planet's surface, for then it follows that the atmospheric pressure and ambient temperature will be in ranges that promote a rich variety of organic reactions.

On the ancient Earth, as today, the simultaneous presence of the three states of matter along terrestrial shorelines provided reaction sites and macroscopic transport for most of the planet's chemicals. The temperature was low enough to confer a substantial lifetime on thermodynamically improbable molecules formed in sunlit waters, yet high enough to give speed to the processes of chemistry, and to the evolution of life. The importance of speed in chemistry and evolution is emphasized by the reflection that a cooler planet than ours, where reaction rates were one-fourth as great, would see its sun burn away from the main sequence of stars before it witnessed intelligent life.

This standard scenario of life's origin has been strengthened greatly by the outcome of laboratory experiments in geochemistry (Fox & Dose, 1972). They show that the assumed primitive molecules of our planet's early atmosphere, if supplied with free energy, could form sugars, amino acids, purines, pyrimidines and other life-related organic substances. More recently, the discovery of numerous interstellar organic molecules by radio astronomers (Robinson, 1976) has all but confirmed that organic molecules of some complexity are of widespread occurrence in the universe. Hence, it is now widely believed that the primordial

* Reprinted from *Origins of Life*, 8(4), 287–98 (1977), with permission of D. Reidel Publishing Company.

125

Earth was the scene of a varied organic chemistry that was rescued from the death of equilibrium by the energy flow from the Sun.

At this point in the development of life on Earth the scene goes briefly out of focus. 'Somehow', it is thought, the first very simple, but living, reproducing cell was formed. The picture then clears as darwinian evolution steps in and drives the progeny of that first cell through inexorable stages of experiment and success to a culmination in intelligent life.

It is the purpose of this chapter to discuss the events that took place between the settling of the Earth's crust about four thousand million years ago and the appearance of the first living organisms a few hundred million years later. The approach is through information theory, for the light that it throws on the probabilities of random formation of molecules that might have been relevant to the origin of life on Earth.

Information in Sequences

Information theory has been developed to assist in the design of communication channels and the codes used with them. As such the theory deals with the mathematical properties of a message, or sequence of symbols, and is not concerned with *meaning*. However, two concessions are made with respect to meaning. If the message is without significance to the receiver, or if it has been received before, it is conceded to contain no information.

The definition of the information content of a message is constructed to meet two intuitive needs. The amount of information should be proportional to message length, and it should increase appropriately with the richness of the alphabet in which it is encoded. The more symbols there are in the alphabet the greater is the number of different messages of length n that can be written in it. The number of permutations of k alphabetic symbols, taken n at a time, is

$$N = k^n \tag{14.1}$$

where N is the number of different possible messages. The information content, H, of a significant message is defined to be

$$H = \log N \tag{14.2}$$

If the logarithm is taken to the base 2 (as always in this chapter), the result is expressed in binary units, or bits.

The probability that a random sequence of n symbols would convey the intended message exactly is only $1/N$. In this sense a significant message is an improbable event. If improbability is taken to be the reciprocal of probability, formula (14.2) states that the amount of information in a message is simply the logarithm of its improbability of occurrence.

An alternative formula for H can be obtained by substituting (14.1) into (14.2), yielding

$$H = \log k^n$$
$$= n \log k \tag{14.3}$$

Clearly, the information content of a message is proportional to its length, and also to the logarithm of the number of symbols in the alphabet it uses. In fact, $\log k$ is the information content per symbol. Because (14.3) is linear in n, the H-values of several messages received in succession (like one longer message) are additive.

The strict validity of these equations requires the condition that each of the k symbols be equally likely to occur anywhere in the sequence. In other words, the code must be nonredundant. This assumption is not seriously violated in the circumstances that will be examined.

That H is a conveniently compact measure of information is seen when formulas (14.1) and (14.2) are applied to a biochemical sequence such as messenger ribonucleic acid (mRNA). Assuming that an average gene comprises a string of $n = 1200$ nuclear bases of $k = 4$ kinds, the number of different genes of that length is, by (14.1), the enormous number

$$N = 4^{1200} = 2^{2400} = 10^{722.5}$$

whereas the information content, using (14.2), is

$$H = \log 2^{2400} = 2400 \text{ bits}$$

a manageably small number.

Apart from purposes of illustration, it is easier to arrive at this result by using a form of eq. (14.3). Since $k = 4$, $\log k = 2$ and therefore

$$H_{\text{RNA}} = 2n$$
$$= 2400 \text{ bits} \tag{14.3a}$$

There are $k = 20$ different amino acids in a modern peptide chain, or protein. Therefore, $\log k = 4.32$ and (14.3) becomes

$$H_{\text{PEP}} = 4.32 \, n \tag{14.3b}$$

for a chain of n amino acids.

An important property of communication channels can now be brought out by comparing the information content of a typical protein with that of the gene that specifies it. Because of the triplet nature of the genetic code the 1200-base gene

Table 14.1. *The information content of various structures*

Structure or sequence	No. of different possible cases or states	Amount of information (bits)
A molecule that is certainly present	1	0.
Toggle switch	2	1.
Nuclear base	4	2.
Amino acid	20	4.32
Combination lock	10^6	20.
Modern tRNA molecule	10^{45}	150.
Random experiments	10^{60}	200.
Gene or protein	$10^{722.5}$	2400.
Virus (50 genes)	$10^{36\,000}$	1.2×10^5
E. coli (2500 genes)	$10^{1\,800\,000}$	6.0×10^6
Man (100 000 genes)	$10^{72\,000\,000}$	2.4×10^8

considered above has 400 codons and therefore is translated into a protein of 400 amino acids. But, by (14.3b), the information content of the protein is only 4.32 × 400 = 1729 bits. About 671 bits have been lost in translation. This feature of genetics is known as the *degeneracy* of the genetic code (64 codons represent only 20 amino acids plus start and stop signals). More physically, it is a degeneracy of the translation process. Clearly, a cell equipped with about 60 different transferase molecules (tRNA) instead of 20, could utilize 60 different amino acids for protein manufacture – all without any change in the basic structure of the mRNA. Degeneracy may have been important for the origin of life, and will be mentioned again.

A useful 'benchmark' to the information content of living organisms is provided by the common bacterium *Escherichia coli*. Assuming it to have 2500 genes, it will encode a total of six million bits. The amounts of information stored in this and other relevant structures are displayed in table 14.1.

At this point it is important to make clear the nature of the argument from information theory as it applies to the origin of the first living organism. We are willing, perhaps, to make the optimistic assumption that the formation of random peptide chains of amino acids was thermodynamically favorable, or at least permitted, thanks to the presence of the necessary organic substances, mineral catalysts and available energy. In other words, random peptides were commonplace.

But a random chain of n amino acids contains no information unless it has significance for the origin of life. If it, alone among such chains, has that significance, its information content will be given by formula (14.2) with N set to the number of different possible chains of length n. (Equally, depending on one's predilection regarding the manner in which life started, the formation of a significant nucleic acid would generate information.)

If life could have started in but one way it would be possible, in principle, to deduce how it began, given only that there is life. Such an origin would therefore have been, in the language of physics, 'nondegenerate'. But if two or more distinct sequences of amino acids (or bases) could have independently triggered the process, then the origin of life would have been a degenerate event. The mere knowledge that life exists would no longer suffice to deduce the actual mode of origin out of the several that were possible. Clearly, life on earth today is a highly degenerate process in that there are millions of different gene strings (species) that spell the one word, 'life'.

The Generation of Information

If the first cell was as complex as *E. coli*, a momentary suggestion made only for illustration, it would have been necessary for our planet's early random chemistry to have generated about six million bits of information by pure chance. The way in which chance experiments can generate information is now examined by looking at a game.

Consider a locked door equipped with a combination lock comprising 20 toggle switches on its outer surface. If all 20 switches are set correctly, the door can be opened by grasping and turning the handle. A person having no knowledge of the correct combination can open the door by making a (long) series of random experiments. A few switch positions are changed at random and the handle is tried. This procedure is repeated until the door opens. The experimenter can now write down the combination by looking at the successful switch positions. Clearly, he has acquired 20 bits of information, 1 bit per switch, by making, probably, about one million random experiments. (There are $N = 2^{20} \approx$ one million different possible switch combinations.)

Here again it is seen that a random sequence (of switch positions) is uninformative as long as it is undistinguished. But as soon as one sequence proves to be significantly different from the other million, it carries 20 bits of information. The average amount of information H_r that can be generated by N random experiments is seen to be

$$H_r = \log N \tag{14.4}$$

and this is merely a specialization of eq. (14.2).

To say that random experiments can *generate* information is perhaps a subjective view not warranted by the facts. It could be argued that the information necessary to open the door is already encoded in the pattern of the electric wiring that connects the switches to the lock mechanism, and that all one need do to acquire that information is to look at the other side of the door! Similarly, it will be true that the sequence of units necessary to the first reproductive structure was already encoded in some deep way in the environment that gave it birth. This may be what the philosophers mean when they say that the potential for life inheres in the very nature of matter.

The Probability of the Origin

In the standard scenario the Earth's fluids were in a continual turmoil of chemical change that produced organic sequences in a random way. All but one of these long molecules were passive. The unique molecule quickly organized most of the carbon in its environment into copies of itself. The information necessary to reproduce had been transferred from the chemical milieu in which it had lain concealed to a molecule capable of recording it. The random chemistry was then replaced by darwinian evolution.

Calculation of the probability that a reproductive chemical was formed by chance requires a knowledge of the rate at which relevant random reactions occurred, and the length of time they continued. Neither figure is well established – especially the first – and no realistic calculation of the information-generating power of the environment can be made. On the other hand, it is possible to set optimistic upper limits to the reaction rate and the available time, and then to derive the maximum possible value of H_r permitted under the assumptions made.

Assuming that the early waters of Earth contained 10^{44} carbon atoms (Suess, 1975), it is optimistic to take the number of amino acid molecules as 10^{43}. If these molecules were linked in random sequences of average length 10, there would be about 10^{42} peptide chains. If the mean lifetime against extension or breakage for the average chain were 10 milliseconds, and the action continued for 500 million years, the total number of peptides formed would be

$$(10^{42} \text{ peptides}) \times (5 \times 10^8 \text{ yr}) \times (3 \times 10^7 \text{ s yr}^{-1})$$
$$\times (100 \text{ reactions s}^{-1} \text{ peptide}^{-1}) = 1.5 \times 10^{60}$$

However, not all of these are different peptides. For the case in which there are 20 kinds of amino acids it can be shown that 98.2% of the 1.5×10^{60} reactions are repeat production of short peptides. Subtracting these, there remain

$$N = 2.75 \times 10^{58}$$

different molecules, and the corresponding information content is, by (14.4),

$$H_r = 194 \text{ bits}$$

If the calculations are made for nucleic acids instead of peptides, the result is substantially the same.

It would seem impossible for the prebiotic Earth to have generated more than about 200 bits of information, an amount that falls short of the 6 million bits in *E. coli* by a factor of 30 000. A natural attempt to save the scenario is to postulate a simpler first cell. However, there is little to be gained through this proposal. An average virus codes about 2% as much information as *E. coli* (120 000 bits) and is not capable of reproducing in an abiotic environment. Rather it must subvert the metabolic machinery of a regular cell for materials, energy and protein synthesis. It is difficult to imagine an independently reproductive cell as simple as a virus (Watson, 1970–1), and even if one can, it helps little to bridge the enormous information gap between chemistry and life.

Parenthetically, it is interesting to note that if the probability of the chance appearance of life on Earth seems remote, there is little comfort to be gained by enlarging the arena to the whole Galaxy. Even if there are 10^9 Earth-like planets in the Milky Way, the potential for random generation of information rises only to 224 bits – less than 0.2% of the content of the average virus.

Even one gene of average length encodes about 2400 bits, so it is not useful to speak of a primitive naked gene that reproduced unless it was so short that it specified a protein of no more than about 33 amino acids. Whether one prefers to think of the first nucleic acid, the first gene, the first protein or the first enzyme as the unique structure that began life, there is the difficulty of visualizing the way so small a molecule could have commanded the environment to its selective reproduction.

If life on Earth had a spontaneous origin, there must have been an intermediate mechanism that was capable of augmenting the information content of one or a few early molecules up to the million-bit level required by the first organism. But before turning to a possible new mechanism, it is worthwhile to consider how darwinian evolution is able to generate information so much faster than random experimentation. The question is philosophically interesting because, as stressed in every good introduction to the subject, the mutations that provide the raw material of evolution occur at random and are made without purpose or goal.

Also, the random experiments, mutations, occur very slowly, about one per generation, and require a great deal of carbon per individual – about 10^{10} atoms. If most of the ocean's carbon resided in cells like *E. coli*, reproducing once per hour for three billion years, the total number of mutations would be only about 10^{47}, less than 10^{-13} of the number allowed for random chemical experiments. Nevertheless darwinian evolution has been able to amass prodigious amounts of information in the world's living species. How this was done can be seen by examining a modified version of the game with the locked door.

Darwinian Generation of Information

Consider now a slightly different kind of door – one that opens a millimeter for each switch that is set correctly. The initial switch positions are random. By noticing the door's response, the uninformed operator can open it very quickly. He selects a switch at random and alters its setting. If the door opens a little, he makes another experiment. Otherwise he returns the switch to its original (correct) position and then goes on to the next experiment. Twenty systematic trials would have opened the door, but about 60 are necessary if they are random, because of needless experimenting with switches already tried.

This door-opening game can be seen to correspond to an extremely simplified model of darwinian evolution if certain identifications are made. The door is the reproducing organism and its openness is a measure of its success. In each experiment (generation) the door is seen in two states – its original state, and its new state after a switch is thrown. These are the two progeny of a divided cell, one normal and the other mutant. Only the 'fitter' of the two states is tolerated by the operator (the environment), which then destroys the other state (offspring) and leaves the former to serve as original state (parent) for the next experiment (generation).

The 20 bits of information required to open the darwinian door have been generated in only about 60 experiments (generations), instead of the million required to open the random door treated earlier. Thus,

$$H_D = cN \tag{14.5}$$

where H_D is the amount of information generated by N darwinian experiments, and c is a factor that equals $\frac{1}{3}$ in this example.

Of course, it is not accurate to assume that every mutation that makes the base sequence of a gene more like that of a superior gene will necessarily itself be an improved gene. In other words, the shortest route to a better gene will not necessarily thread a series of monotonically improving genes. Therefore the formula for H_D gives an upper limit rather than a precise estimate of the amount of information generated in N experiments. Furthermore, the factor $\frac{1}{3}$ is appropriate only to the game described. It will be smaller for longer genes carrying more than 20 bits, for mutation rates less than the 50% used in the game, for larger populations (because of duplicate mutations) and for nearly perfected genes that are harder to improve. Unpublished computer simulations suggest a c-value in the range 10^{-2} to 10^{-6} for plausible simple model organisms.

Even the lower values of c have little impact on the disparity between H_D and H_r for interesting values of N. When $N = 10^{40}$, $H_D = 10^{34}$ and $H_r = 132$. The darwinian organism acts like a machine for generating information. Its special function is to copy all its genes, including those carrying random alterations to the message. All new messages are then subjected to environmental scrutiny. It is immaterial that only rarely does a new message pass the test. Once approved, it is copied at an exponentially increasing rate and, for a time, becomes the 'standard' message that underlies future attempts to encode even more information about the environment's tolerance for life. This process works because the reproductive power of a population of organisms exceeds the environmental culling that takes place between generations. The cost imposed by genetic experimentation is paid out of surplus reproduction. Meanwhile mutations that are not rejected add information to the genes at the darwinian rate.

An Intermediate Mechanism

Any proposed new mechanism for generating information faster than possible by pure chance faces a fundamental *logical* difficulty if it is embodied in an integrated structure. Any structure that reproduces at a rate that outruns decay processes will undergo darwinian evolution, and for that very reason will be a darwinian organism. This dilemma can be avoided by dropping the tacit assumption that reproduction is possible only to an integrated structure such as a cell or other living organism. If an amorphous *community* of free molecules could reproduce, it could also evolve in a proto-darwinian way, but would not be living, because it would contain no structure that independently reproduced itself.

Although a cell such as *E. coli* would not reproduce in today's environment if it lost its wall, there seems to be no impediment to the general idea that the contents of a broken cell (along with the wall fragments), when poured into a benign environment, might still carry on the biochemical reactions characteristic of reproduction (Horowitz, quoted in Margulis, 1970). If that happened, there would be a multiplication of the populations of biochemicals at the expense of the simpler organic precursors in the region. Such a system would contain no living entity, but certain molecules in it would make copies of each other even though no molecule reproduced itself. It must now be asked whether one can postulate a chemically reproductive community containing less than 200 bits of information without forsaking plausibility.

A Reproductive Chemical Community

An important requirement for plausibility in any proposal for the origin of life is the recalculation of H_r on a less wildly optimistic basis. To this end the assumed concentration of amino acids (or nucleotides) is reduced (by the factor 10^3) to 10^{40} in all the oceans; the reacting mass is reduced from the entire hydrosphere to those waters shallow enough to permit sunlight to reach the bottom, a factor of 100; and the lifetime of an average molecule is increased to 1000 s, a factor of 10^5. The product of these factors is 10^{10} and the lost information entailed by their use is 33 bits. Thus a more plausible value for the amount of randomly generated information is

$$H_r = 161 \text{ bits}$$

The idea of a reproductive chemical community can be made concrete by adopting the suggestion of Crick *et al.* (1976) that early protein synthesis took place without the use of even a simplified ribosome. Their scheme achieves translation of a primitive genetic code to protein by means of transfer RNA molecules (tRNA) each carrying a 7-base anticodon loop at its 5' end and a specific amino acid recognition group at the 3' end. An attempt will now be made to calculate the minimum amount of information required to create the necessary molecules.

In its simplest and most cogent form the new mechanism recognizes the four amino acids glycine, asparagine, serine and aspartic acid, whose modern codons occur in the bottom right-hand corner of the codon table. Consequently, four tRNAs are involved in the translation of a mRNA molecule. Because of the degeneracy of the 3-base codon only one bit is required for each base in the mRNA, and therefore also for each of the three variable bases in the 7-base anticodon. The other four fixed anticodon bases are nondegenerate and represent two bits each. Thus 11 bits are required for each of the four 7-base anticodons.

The problem of amino acid recognition is not dealt with by Crick *et al.* Presumably each tRNA forms a recognition cavity at its 3' end. Allowing eight bases for the cavity would call for 16 bits for each tRNA. However, the occurrence of considerable degeneracy, especially in primitive structures, is widespread in the biochemical world (Hasegawa & Yano, 1975) and it may not be unreasonable to reckon 12 bits per cavity. If functional tRNAs could be this simple, they would embody only 23 bits each.

Only one molecule remains, the mRNA, and it will be assigned the remaining information. Subtracting 4×23 from the allowed 161 bits leaves 69 bits for an mRNA molecule containing 23 3-bit codons. Without further appeal to degeneracy it is now entirely proper to demand that the mRNA specify the best nonspecific RNA replicating enzyme possible out of all peptides composed of 23 amino acids, each chosen from the set of four.

At this stage in the development of the idea of a reproductive chemical community, it would be all too easy to conclude that the replicase would form copies of all five RNAs while they were busy making more replicases. Indeed, the formation of the first replicase *starts* promptly because the principle of additivity of information used in this treatment applies only to the case where all the specified molecules are made at the same time, in the same place and in specified relationship to each other. That is, the 161 bits of information include specification of the correct initial juxtaposition of the five molecules to *begin* translation. But translation will not be completed unless the used tRNAs that drift away from the mRNA after the formation of each peptide bond are brought back *again and again* by Brownian diffusion until all 22 bonds have been made. Furthermore, all this must happen during the postulated 1000 second lifetime of the molecules. To ensure completion of the first replicase, it may be necessary to suppose that the crucial molecules were trapped in a micrometer-sized interstice between particles of clay or other material. After a substantial population of each of the six molecules had been generated the system would become secure against diffusion, and would no longer require, or benefit from, confinement in a small volume.

Proto-darwinian Evolution

In the modern cell mRNA codons can be translated into amino acids at the rate of 40 s^{-1}, with an error ratio below one in a thousand. Replication of nucleic acids is about equally fast, and nearly a million times more accurate (Watson, 1970–2). In the first reproductive chemical community envisaged here the population would double in every generation if about two reasonably accurate sets of six molecules were produced every 1000 seconds, per mRNA. Thus, a high mutation rate combined with sluggish replication and translation would be permitted, as well as expected. If such a system generated information at the rate of 0.1 bit per mutation ($c = 0.1$ in formula 14.5), only a few hours would be required to double the original investment of 161 bits. Thus, the onset of reproduction and proto-darwinian evolution in a randomly generated system that might have taken hundreds of millions of years to arise marks an exquisitely critical point in its history.

In a proto-darwinian community there are no organisms to compete with each other. The information necessary to reproduction is dispersed in several free RNA molecules. (The central dogma of darwinian genetics, that information flows from nucleic acid to protein, but not along the reverse path, is already visible in the fact that the peptide molecule, replicase, is not an essential element of the reproductive set.) Tolerable errors in translation or replication lead not to new organisms but to new chemical pathways, and these compete. Because of the slowness of long-range molecular diffusion, competition is mostly a local matter. A novel pathway can survive for a time in its own region even though it is not the best in the pool. But in the long run the pool evolves as a whole.

Each mRNA molecule acts as a centre of translation, accepting all appropriate tRNAs that drift into contact with it. Thus, each replicase is assembled by random tRNAs in the community surrounding the mRNA. It is noteworthy that the system of Crick et al., as developed here, posits more information in the tRNAs than in the mRNA. If the five RNAs are all regarded as genes, it could be said that proto-darwinian reproduction is hypersexual in the sense that nucleic acids mix even more freely than in sexually pairing organisms.

Assuming that the reproductive chemicals eventually leak into new niches and evolve complexity and diversity, there would soon be a variety of specific pathways plied by separate communities of chemicals of ever-increasing molecular weight and specificity. However, proto-darwinian systems lack one important feature of darwinian populations – safeguards against hybridization. Two divergently evolved chemical communities spilling into a common pool would unavoidably hybridize if their pathways contained any common segments, as would seem very likely. For example, if there had been divergent evolution of genetic codes the indiscriminate mixing of nucleic acids would thwart the interlocking specificities so laboriously built up over their histories and would waste resources prodigiously. This difficulty would dog the path of proto-darwinian evolution until some means of collecting and sheltering a reproductive set of chemicals was evolved. Perhaps the darwinian organism was evolution's answer to the problem of hybridization.

It is easy to suggest ways in which a reproductive set of chemicals might clump together in relative isolation from the rest of the community. The difficulty is to have the isolation partial in just the way that permits essential precursor molecules to enter the enclave but prohibits the loss of genetic material until it can be discharged in a self-reproductive clump. *That this partial isolation is not easy to specify is just the problem of the origin of life.* It probably required a million bits of information, and it is the burden of this chapter that that information might have been generated by the proto-darwinian evolution of a reproductive community that began with less than 200 bits of randomly generated information.

The shortcomings of this scheme to start some kind of rapid information-generating process in the prebiotic soup are too obvious to ignore. The broth is speciously thick and the prospects of the first community are precarious in the extreme. It would be helpful to know more about the catalytic powers of short peptides, especially those containing only two kinds of amino acids. Such enzymes could be assembled by two tRNAs. If the RNA replicases turn out to be highly degenerate, less information will be needed for their formation. These and other possibilities remain to reduce the improbability of the chance occurrence of a reproducing system.

Conclusions

Improbable structures can be formed by random trials if the latter are sufficiently numerous. Information theory simplifies the task of separating the possible from the impossible by reducing structural complexity and experimental prodigality alike to a common informational measure expressed in bits.

The calculation of the information-generating power of the Earth's primitive hydrosphere offered here is neither precise nor definitive. Rather it is suggestive that there is an enormous information gap between the products of a random chemistry and the simplest imaginable reproducing organism.

It seems futile to force darwinian evolution backwards through simpler and simpler organisms to one whose structure could have been the outcome of random trials. Instead it is proposed that special molecules that arose by chance formed a reproductive community of sufficient vigor to start a proto-darwinian evolution that dominated its development.

Proto-darwinian evolution will have been significant for the origin of life if at least one reproductive chemical community can be specified by not more than 200 bits of information, and does not lie in an evolutionary cul-de-sac.

If the 200-bit figure is seriously in error, it is too large. If the true figure is less than half this upper limit, it will probably be necessary to discover information-generating mechanisms beyond those discussed here. Alternatively, it would be encouraging to discover that enzymes are highly degenerate molecules that economize on information.

Note Added in Proof

The prospects of the reproductive chemical community described in pp. 134–5 become more promising when it is noted that the system is slightly degenerate. If the specifications of the five molecules are thought of as five words in a 161 bit message, it is seen that there will be 5! permutations of word order, and that these do not change the meaning. Moreover, there are 4! possible sets of associations between the anticodons and the amino acid recognition sites of the four tRNAs. Although each such set would require a different mRNA to specify the replicase molecule, the appropriate messenger can always be encoded by the 69 bits allotted to it. Consequently, the postulated system would arise 4! × 5! = 2880 times at diverse places in the hydrosphere. Only one of these systems need have propagative success to start proto-darwinian evolution.

References

CRICK, F. H. C., BRENNER, S., KLUG, A. & PIECZENIK, G. (1976). *Origins of Life*, 7, 389.

FOX, S. W. & DOSE, K. (1972). *Molecular Evolution and the Origin of Life*, chapter 4. San Francisco: W. H. Freeman.

HASEGAWA, M. & YANO, T. (1975). *Origins of Life*, 6, 219.

MARGULIS, L. (1970). *Origins of Life*, p. 311. New York: Gordon and Breach.

ROBINSON, B. J. (1976). *Proc. Astron. Soc. Aust.* 3, 12.

SUESS, H. E. (1975). *Origins of Life*, 6, 9.

WATSON, J. D. (1970–1). *Molecular Biology of the Gene*, p. 503. New York: Benjamin.

WATSON, J. D. (1970–2). *Molecular Biology of the Gene*, pp. 297, 368. New York: Benjamin.

15

The RNA World: Life before DNA and Protein*

GERALD F. JOYCE

Introduction

All of the life that is known, all organisms that exist on Earth today or are known to have existed on Earth in the past, are of the same life form: a life form based on DNA and protein. It does not necessarily have to be that way. Why not have two competing life forms on this planet? Why not have biology as we know it and some other biology that occupies its own distinct niche? Yet that is not how evolution has played out. From microbes living on the surface of antarctic ice to tube worms lying near the deep-sea hydrothermal vents, all known organisms on this planet are of the same biology.

Looking at the single known biology on Earth, it is clear that this biology could not have simply sprung forth from the primordial soup. The biological system that is the basis for all known life is far too complicated to have arisen spontaneously. This brings us to the notion that something else, something simpler, must have preceded life based on DNA and protein. One suggestion that has gained considerable acceptance over the past decade is that DNA and protein-based life was preceded by RNA-based life in a period referred to as the 'RNA world'.

Even an RNA-based life form would have been fairly complicated – not as complicated as our own DNA- and protein-based life form – but far too complicated, according to prevailing scientific thinking, to have arisen spontaneously from the primordial soup. Thus, it has been argued that something else must have preceded RNA-based life, or even that there was a succession of life forms leading from the primordial soup to RNA-based life. The experimental evidence to support this conjecture is not strong because, after all, the origin of life was a

* Edited transcript of a public lecture entitled 'The Dawn of Biology Current Views Concerning the Origins of Life', presented at the Scripps Institution of Oceanography, La Jolla, California, 16 May 1991.

historical event that left no direct physical record. However, based on indirect evidence in both the geologic record and the phylogenetic record of evolutionary history on earth, it is possible to reconstruct a rough picture of what life was like before DNA and protein.

What is Life?

It is useful, at the outset, to consider what is meant by the word 'life'. This word has a vague popular meaning, making it difficult to provide a rigorous scientific definition that will satisfy all audiences. The popular definition of 'life' might be stated simply as: 'that which is squishy'. Life, after all, is protoplasmic and cellular. It is made up of cells and organic stuff and is undeniably squishy. A more mechanistic popular definition might be: 'life is that which eats and procreates'. In a very broad sense, living organisms turn food into offspring. They metabolize food and use the energy derived from the food to produce offspring, that is, to produce more life. Among biologists and biochemists a current working definition of 'life' is: 'a self-sustained chemical system capable of undergoing darwinian evolution'. This is the definition that I shall adopt for the following discussion. 'Life' is that which evolves in a darwinian sense.

Darwinian evolution occurs as a result of three physical processes: amplification, mutation and selection. Amplification involves the replication of a prototype, or, more precisely, the replication of a genetic description of a prototype. Mutation is a process that introduces variation during replication of the prototype. Selection involves choosing among the various replicates to establish a new prototype, which is then used to begin another round of amplification, mutation and selection.

How does terrestrial biology embody these three processes? It relies on DNA to provide a genetic description of the prototype. The DNA contains instructions describing, in effect, how to build and operate the organism. These instructions are copied from DNA to RNA. The RNA then acts as a messenger to carry the instructions to a complex cellular apparatus, the ribosome, where the instructions are interpreted to produce proteins. The resulting assemblage of functional proteins might be called the prototype or, more formally, the 'phenotype' of the organism.

Biology carries out amplification by replicating the genetic description of phenotype, that is, by replicating the DNA. Mutations occur during the replication process, so that the DNA copies resemble, but are not identical with, the parental DNA. No two copies are exactly alike. Mutations that exist in the DNA copies are interpreted by the ribosome as altered instructions for the production of proteins. This results in somewhat altered proteins that may have altered function.

Some functional variations will be more useful than others, and it is the variations that are most favorable, together with the DNA that describes them, that are selected to begin the next round of amplification, mutation, and selection. And so it goes, round after round, generation after generation. The power of darwinian evolution, and the success of life on earth, are attributable to the very large number of repetitions of this cycle that can occur. Biology on Earth has undergone trillions of rounds of amplification, mutation and selection. These events, played out on a global scale, constitute the natural history of our planet.

However, as alluded to previously, this is all too complicated if one is thinking about the prebiotic Earth. It is not so difficult to imagine how an instruction in DNA could be copied over to an RNA messenger. But it is very hard to imagine how that message could be translated into protein without the aid of a complex biochemical apparatus such as the ribosome. If one is considering a time prior to the origins of life on Earth, then a translation apparatus would not yet have been invented. It would require a great many rounds of darwinian evolution for a functional entity as complicated as a ribosome to develop.

Life Based on RNA

So how does the game get started? What is the solution to what is often referred to as 'the chicken-and-egg problem'; the egg being the genetic instructions contained within DNA and the chicken being an expression of phenotype at the level of protein function? An important insight that has taken hold in recent years stems from the observation that, like proteins, RNA can have complex function. Biological phenotype derives from the function of cellular enzymes, and these enzymes may be comprised of either protein or RNA. A discovery that revolutionized our understanding of biology, for which Thomas Cech and Sidney Altman shared the 1989 Nobel Prize in Chemistry, is that RNA can be both a carrier of genetic instructions and an agent that exhibits enzymatic function (Kruger et al., 1982; Guerrier-Takada et al., 1983).

Why not, therefore, have a life form that is based solely on RNA, in which RNA is at once both the instructional molecule, the genotype, and the functional molecule, the phenotype? RNA as an instructional molecule can be amplified, subject to mutational error, to produce progeny copies of variable composition. RNA as a functional molecule can be subject to a selection process, such that those individuals that are best able to solve problems imposed by the environment are chosen as the prototypes to begin the next round of amplification, mutation and selection.

RNA is a polymer made up of subunits, termed 'nucleotides'. The subunits are of four types: adenosine (A), guanosine (G), cytosine (C) and uracil (U). It is the specific ordering of the subunits within the polymer, for example A–U–G–U–C–A–A–G–U . . . , that constitutes the genetic information. An RNA molecule can assume a well-defined structure in water, based on the particular ordering of the subunits that it contains. This structure, in turn, causes the molecule to exhibit particular functional properties.

What is the evidence that an RNA-based life form actually existed on this planet prior to the emergence of DNA and protein-based life? First, it is known that RNA can function as a genetic molecule. There are a number of viruses in existence today that utilize RNA, rather than DNA, as their genetic material. There is no known example of a free-standing RNA-based organism, which would constitute an extant RNA-based life form. All of the known RNA viruses are parasites of DNA- and protein-based organisms and thus must be considered part of our own biology. However, the existence of RNA viruses demonstrates that RNA genomes can exist. A second piece of evidence comes from the work of Leslie Orgel and colleagues, who have shown that, in a purely chemical system, an RNA molecule can be made to copy itself (von Kiedrowski, 1986; Zielinski & Orgel, 1987). The copying process is intolerant of mutations, and thus these RNA molecules do not begin to evolve. But, again, this is a demonstration of the principle that RNA can be a carrier of amplifiable genetic information. A third piece of evidence favoring the possibility of an RNA-based life form is the discovery that RNA can function as an enzyme. There are now many known examples of RNA enzymes in biology (for reviews see Cech, 1987, 1993). This establishes the fact that RNA can be a functional molecule as well as a genetic molecule, meaning that it has the wherewithal to provide the chemical basis for darwinian evolution. This is not proof, however, that such a situation actually existed.

If it really did happen, if there was a time when life on Earth was based on RNA before it gave way to DNA and proteins, then one might expect to see remnants of the prior RNA-based life form within the succeeding DNA- and protein-based life form. What, then, is the role of RNA in our present life form? It seems to be involved in just about everything, especially as concerns the most central, most highly conserved, most primitive aspects of cellular function. RNA is a messenger, carrying genetic instructions from DNA to the protein-synthesizing machinery. RNA is an integral part of the protein-synthesizing machinery itself, drawing in the protein subunits in response to genetic instructions and carrying out the process by which the subunits are joined to form mature proteins. RNA is also involved in editing and splicing various bits of genetic information, to properly arrange the genetic instructions prior to translation. RNA is even needed for the replication of DNA. When DNA is copied, the process is initiated by the production of a partial RNA copy, which is then extended to give DNA. In summary, RNA is involved in nearly all of the informational processes of the cell.

What about function? Except for the few oddball RNA enzymes that have been discovered over the past decade, the bulk of cellular function is, after all, carried out by proteins. On the other hand, a close look at the protein enzymes reveals that over half of them, and nearly all of the protein enzymes that are involved in the most fundamental aspects of cellular metabolism, rely on 'coenzymes.' Coenzymes do more than assist proteins in carrying out biochemical reactions; they play a crucial role in the mechanism of the reaction. Remarkably, almost all of the known coenzymes contain components of RNA. They typically consist of one or two nucleotides, together with some chemical attachment. Thus, at the level of function, even protein-based function, the components of RNA are found to be intimately involved.

Finally, consider the two main differences between RNA and DNA: (1) RNA contains the sugar component ribose, while DNA contains the sugar deoxyribose; (2) one of the subunits of RNA is uridine, while the comparable subunit of DNA is methyluridine or thymidine. The way in which biological organisms generate these two different forms is to first produce the RNA version and then, as a last step, convert the RNA version to the DNA version. For example, the deoxyribose sugar of DNA is first produced as ribose, which is incorporated into nucleotide subunits and, at the last step, modified to give deoxyribose. Why not have it the other way around? Why not produce deoxyribose first and ultimately convert it to ribose? One explanation is that RNA existed prior to DNA, so that a means of producing ribose was already in place before there was a need to produce deoxyribose. Similarly, the thymidine subunit of DNA is first produced as uridine, which is ultimately converted to thymidine, again suggesting the primacy of RNA.

If we accept the notion that RNA-based life preceded DNA- and protein-based life, then it is reasonable to wonder when the RNA world first came into existence and when it gave way to DNA and protein. Both the genetic and functional properties of RNA require that it be dissolved in water. Thus, until the newly formed Earth had cooled to the point that liquid water was available, it would not have been possible for RNA-based life to exist. Attempting to set the time frame for the disappearance of the RNA world, one can search the geologic and phylogenetic record for the earliest evidence of DNA- and protein-based life. Between these two endpoints lies the window of opportunity for the existence of the RNA world (Joyce, 1991).

The Prebiotic Earth

The story begins roughly 4.6 billion years ago, generally agreed to be about the time when our solar system was formed. This number is based on isotope dating of lunar rocks returned from the Apollo missions and of fallen asteroids. Prior to 4.6 billion years ago the pre-solar system consisted of a swirling cloud of gas and dust, collapsing under its own gravitational forces. At the center of this giant cloud, the Sun began to form. At varying distances from the center, matter condensed to form planetesimals – solid objects only a few kilometers in diameter. These in turn coalesced to form planets, although the details of this process are somewhat unclear. In any case, within 100 million years, roughly 4.5 billion years before the present, the formation of the Earth was largely complete (Stevenson, 1983).

Over the next 300 million years, from 4.5 to 4.2 billion years ago, the young Earth continued to accrete material, picking up stray planetesimals and debris that lay in its orbital path. As this material impacted the Earth, its kinetic energy was transferred to the planet in the form of buried heat. As a counterbalance, the Earth underwent convective cooling, causing much of the buried heat to be radiated back to space. Eventually, however, the Earth reached a size at which convection could no longer keep pace with the amount of heat being buried, and planetary temperatures began to rise. These conditions made it impossible for liquid water to exist, which in turn made it impossible for RNA-based life to exist.

When was it first possible that there was liquid water on Earth? There are two ways of looking at this problem. On the one hand, suppose that water was present from the time of initial accretion. Then it would have been especially difficult for the planet to maintain a moderate temperature, because water vapor in the atmosphere would act to promote a 'greenhouse effect'. Water vapor, like carbon dioxide, is a greenhouse gas that reflects some of the Earth's radiated heat back to the surface. This would diminish the amount of buried heat that could be radiated to space, resulting in increased surface temperatures, which in turn would cause still more water to evaporate and enter the atmosphere. The culmination of this positive feedback loop would be a runaway greenhouse effect, causing all of the surface water to evaporate and the Earth to be covered by a global magma ocean (Kasting, 1988; Zahnle et al., 1988). Thus, if water was present from the beginning, the dawn of the RNA world could not have occurred until after the Earth's crust had cooled to the point that liquid water could be present again.

On the other hand, suppose that water was not present from the beginning and instead was delivered to the planet at a later time by impacting comets (Chyba, 1987). If water did not arrive until later, then there would not have been a runaway greenhouse effect. Of course, until the water did arrive, RNA-based life would have been impossible. Thus, in either case, the RNA world could not have arisen until some time after the Earth's orbit had been cleared of debris, perhaps 4.2 billion years ago.

However, there is another problem, first pointed out by Maher & Stevenson (1988). Even after planetary accretion was complete, meteors and asteroids continued to intersect the Earth's orbit, occasionally striking the surface. The effect of an impacting meteor or asteroid can be devastating. For example, there was a major impact event at the Cretaceous–Tertiary boundary, 65 million years ago, that is thought to have been responsible for the extinction of the dinosaurs (Alvarez et al., 1980). This so-called 'K–T impactor' is believed to have been an asteroid roughly 50–75 km in diameter that produced a crater about 200–300 km across, spewing vast amounts of vaporized rock into the atmosphere (Sharpton et al., 1992, 1993). This had a profound effect on global climate, and turned out to be devastating for the dinosaurs and other organisms that were unable to evolve quickly enough to compensate for the changes. The K–T impactor was a rare event that occurred recently on the geologic timescale. During the early history of the Earth, however, such events are thought to have been much more common. Asteroids the size of the K–T impactor are thought to have hit the earth about once every 50 000 years at a time 4.2 billion years before the present (Maher & Stevenson, 1988). These events became progressively less frequent over the next half-billion years. But at the time when liquid water first became available and RNA-based life might have just begun to gain a foothold, it would not have been long before a devastating impact event occurred.

What is devastating for a dinosaur may not be so devastating for a microbe or an RNA-based organism. But there also were less frequent, though truly massive, impact events that would have made the K–T event seem like a summer hailstorm. On the basis of extrapolation from the lunar impact record, and taking into account the larger cross-sectional area and greater gravitational pull of the Earth, Maher and Stevenson estimate that massive impact events were occurring with distressing frequency on the early Earth. Impacting bodies having a diameter of 250 km or greater, producing a crater at least 850 km across, were occurring roughly every million years 4.2 billion years ago. An event of this magnitude would be expected to completely sterilize the Earth. Such an event would have been devastating to RNA.

Imagine you are an RNA-based life form, just beginning to evolve into something interesting, when along comes one of these devastating events. Maher and Stevenson appropriately term this phenomenon 'impact frustration' because what could be more frustrating to life? It is important to note that the models of impact frustration must be taken qualitatively. The data concerning the cratering history of the moon are not as complete as one would like. Furthermore, the estimated interval of one million years between global sterilizing events represents an average. There may have been intervals that were considerably longer and allowed life to evolve adaptive countermeasures to survive the next big one. Perhaps life could survive at the bottom of the ocean near the deep-sea hydrothermal vents or developed the ability to enter a protected dormant state until the environmental upheaval had subsided.

In any case, as time went on, such devastating events became progressively less frequent, so that by 4.0 billion years ago they were occurring 'only' about once every ten million years. That may have provided enough time for life to evolve an effective survival strategy. By 3.8 billion years ago, massive impacts were becoming quite rare, occurring perhaps every hundred million years. This is the time, from 4.0 to 3.8 billion years ago, that RNA-based life would have enjoyed its first reasonable opportunity for survival.

Life before RNA

Just because environmental conditions made it possible for RNA to exist does not mean that in fact an RNA-based life form was present. One must consider the three chemical components of RNA: the ribose sugar, the phosphate connector and the nucleotide base (A, G, C or U), and ask whether these components would have been available on the primitive Earth. This is a challenging problem in prebiotic chemistry, but one that has largely been solved. For example, the chemistry needed to produce ribose under plausible prebiotic conditions is reasonably well understood. The availability of phosphate, while a bit more problematic, does not seem to be an insurmountable problem. Of the four nucleotide bases, A and G are expected to have formed quite readily, while C and U would probably have been present in far lower quantities. Thus, it appears that all of the components of RNA would have been present, at least to some extent, on the prebiotic Earth.

A far more difficult problem is one of specificity: explaining why the components of RNA should be assembled to the exclusion of other closely related compounds. The chemistry that produces ribose would be expected to yield many other sugars as well. The chemistry that leads to A and G would provide a variety of related molecules. Attachment of both the phosphate and nucleotide base to ribose would be complicated by hundreds of side-reactions, making it difficult to see how RNA could stand out in the crowd.

Perhaps there were special conditions, in at least one locale on the primitive Earth, that allowed preferential synthesis of ribose over all other sugars, and of A, G, C and U over all other related compounds. For example, Albert Eschenmoser and colleagues have shown that there is a favored route to ribose, provided one begins with the appropriate set of starting materials (Müller *et al.*,

1990). But even if there was a pure solution of all the components of RNA, there remains the difficult task of properly assembling these components to form RNA. Again, one is forced to appeal to a special set of conditions that would allow all of the pieces to come together in just the right way.

There is another problem with RNA, a problem that cannot be resolved by appealing to some special set of conditions. RNA, like most biological molecules, has a handedness. The ribose sugar of RNA can exist in either a left-handed or right-handed form. All of the ribose on the prebiotic Earth would have existed as an equal mixture of the two forms. By a quirk of chemistry, it turns out that the left- and right-handed versions of ribose are excellent mimics of each other, so much so that they spoil each other's ability to replicate (Joyce et al., 1984). Without replication there is no evolution, and without evolution there is no way to devise a biochemical solution to the handedness problem. Biological organisms have solved the problem by ensuring that all of the ribose (and later deoxyribose) that they utilize is of the right-handed form. Biological enzymes themselves have a handedness and are able to distinguish between the left- and right-handed version of ribose when producing RNA. But handedness is a property of life, so that exclusion of the 'wrong'-handed form of ribose would not have been possible until after life had originated.

Faced with the difficulty of assembling the components of RNA under prebiotic conditions, and especially with the problem of handedness, many scientists have come to the conclusion that life did not begin with RNA. What might have come before RNA is open to conjecture. It has been suggested that RNA was preceded by a molecule that lacked handedness, or at least did not face the problem of one hand inhibiting the other (Joyce et al., 1987). It is difficult to assess the plausibility of these theories. The experimental evidence for the RNA world is already scanty and largely circumstantial; the evidence as to what came before RNA is virtually non-existent.

In summary, although it may have been possible for an RNA-based life form to exist on the Earth roughly 4.0 billion years ago, some other life form must have come before RNA. We do not know the chemical nature of that preceding life form or, in fact, whether there were several successive life forms that preceded RNA. This is at present a highly active area of research.

The Antiquity of DNA and Protein

In order to gauge the time of transition from RNA- to DNA- and protein-based life, one can look to the fossil record for the earliest evidence of DNA- and protein-based organisms. In this regard, stromatolites, which are sedimentary rocks of biogenic origin, have provided some of the oldest evidence for life on earth. Stromatolites can be thought of as 'living rocks'. On their surface exists a complex community of microbial life, including numerous forms of bacteria and fungi. Over time, these organisms trap sediment and organic debris, eventually polluting their local environment. The organisms then tend to migrate upward through the debris to re-establish a surficial community where they can flourish again. This cycle of growth, pollution and migration is repeated many times, ultimately giving rise to stromatolites, which have a characteristic laminated appearance.

Stromatolites are remarkable not only for the complex series of biological events that leads to their formation, but also for the fact that they appear virtually unchanged throughout the geologic record over the past 3.5 billion years. Modern stromatolites, such as those being formed on the west coast of Baja California or in Shark's Bay, Western Australia, are extraordinarily similar in appearance to specimens found in South Africa that have been dated as 2.3 billion years old. This similarity applies not only to their gross morphology, but also to their fine structure, such as the tracks left by migrating organisms and the remnants of cellular debris (Walter, 1983). The continuity of life, as represented by the stromatolites, extends deep into the geologic record. The oldest known stromatolites, found in Western Australia, have been dated at 3.56 ± 0.03 billion years before the present.

Modern stromatolites are produced by DNA- and protein-based organisms. One might imagine that the very oldest stromatolites were produced by RNA-based organisms. But it would require an astonishing coincidence for the RNA version and DNA/protein version of stromatolites to be nearly identical. Stromatolites reflect the physical and behavioral properties of an entire community of organisms. Surely RNA-based life would have done things somewhat differently than DNA- and protein-based life. Thus, the most parsimonious hypothesis is that all of the known stromatolites derive from DNA- and protein-based organisms. This places the boundary for the transition from RNA to DNA and proteins at a time prior to 3.56 billion years ago.

The oldest direct fossil evidence of life on Earth comes from the work of William Schopf and colleagues (Schopf & Packer, 1987; Schopf, 1993). They have obtained microfossils, i.e. fossils of microscopic organisms, dated at 3.46 billion years before the present. These fossil organisms are very similar in appearance to more recent examples of cyanobacteria that occur throughout the geologic record. Again, the continuity of form seen in microfossils over the past 3.5 billion years argues for the continued existence of DNA- and protein-based life over that same period of time. The time frame for the RNA world seems to be constrained to the half-billion year interval between 3.5 and 4.0 billion years ago.

The geologic record extends a bit further than the time of the oldest known stromatolites and microfossils. The oldest well-characterized rocks are 3.77 billion years old, found in the Isua region of southwestern Greenland. These rocks have undergone metamorphosis at extremely high temperature and as a result are not expected to contain fossil evidence of life. However, the organic carbon in these rocks is very slightly enriched in the isotope ^{12}C relative to ^{13}C, and this observation has been taken by some as indirect evidence of life (Schidlowski *et al.*, 1983; Schidlowski, 1988).

Biological organisms that fix carbon, i.e. convert carbon dioxide to sugar, do so with the help of protein enzymes. These enzymes tend to discriminate among the various isotopes of carbon, preferentially incorporating ^{12}C while excluding ^{13}C. Organic debris of biogenic origin tends to be slightly enriched in ^{12}C, as is true, for example, of the material obtained from the 3.56 billion year old stromatolites discussed above. The very slight ^{12}C enrichment of the 3.77 billion year old rocks from Greenland is a soft call at best. It has been argued that the reason that the enrichment is so slight is because intense metamorphosis has allowed partial re-equilibration of the carbon isotopes (Schidlowski, 1988). In effect, this is arguing that the carbon isotope evidence for life is lacking, but is lacking in just the way one would expect if life had been present. It is fair to say that there is no substantive claim for life, let alone DNA- and protein-based life, older than 3.56 billion years.

A New Approach

There is another approach to the problem of the existence of the RNA world. If one believes that an RNA-based life form is possible, then why not make one in the laboratory? This approach is not meant to diminish the importance of scientific issues such as: Where did ribose come from? Why was ribose the preferred sugar? Where did the nucleotide bases come from? Why were particular bases chosen for RNA? How were the components of RNA joined together? How was the handedness problem resolved? When did RNA first begin to replicate? How did it survive massive impact events? How did life make the transition from RNA to DNA and protein? But a research biochemist knows how to obtain the components of RNA: they can be bought from a chemical supply house! These components are available as pure compounds having only the proper handedness. They can be assembled in the laboratory to produce RNA.

The challenge is to devise RNA molecules that have the ability to direct their own replication. Replication should be made to occur with occasional mutations, so that the progeny copies resemble, but are not identical with, their parents.

Selection would be expected to occur automatically, favoring those molecules that replicate most rapidly under the prevailing reaction conditions.

Progress is occurring along these lines. It is now possible to amplify, mutate, and select large populations of RNA molecules in the laboratory (Joyce, 1989). So far, these RNA molecules have not demonstrated the ability to replicate themselves; it is up to the experimenter to carry out RNA amplification. But RNA evolution can be made to occur, leading to the development of new and interesting RNAs whose functional properties conform to the demands of the experimenter (Beaudry & Joyce, 1992; Bartel & Szostak, 1993; Lehman & Joyce, 1993). This laboratory process cannot be called 'life' because it is not a *self-sustained* chemical system capable of undergoing darwinian evolution. It requires the active intervention of the experimenter. However, it is probably only a matter of time, to be measured in years rather than decades, before a self-sustained RNA-based evolving system can be demonstrated in the laboratory. This would be a case in which a DNA- and protein-based life form, namely a human biochemist, gives rise to an RNA-based life form, an interesting reversal of the sequence of events that occurred during the early history of life on Earth.

References

ALVAREZ, W., ALVAREZ, L. W., ASARO, F. & MICHEL, H. V. (1980). *Science*, **208**, 1095–108.

BARTEL, D. P. & SZOSTAK, J. W. (1993). *Science*, **261**, 1411–18.

BEAUDRY, A. A. & JOYCE, G. F. (1992). *Science*, **257**, 635–41.

CECH, T. R. (1987). *Science*, **236**, 1532–9.

CECH, T. R. (1993). In *The RNA World*, ed. R. F. Gesteland & J. F. Atkins, pp. 239–69. Cold Spring Harbor, NY: Cold Spring Harbor Laboratory Press.

CHYBA, C. F. (1987). *Nature*, **330**, 632–5.

GUERRIER-TAKADA, C., GARDINER, K., MARSH, T., PACE, N. & ALTMAN, S. (1983). *Cell*, **35**, 849–57.

JOYCE, G. F. (1989). *Gene*, **82**, 83–7.

JOYCE, G. F. (1991). *New Biologist*, **3**, 399–407.

JOYCE, G. F., SCHWARTZ, A. W., MILLER, S. L. & ORGEL, L. E. (1987). *Proc. Natl. Acad. Sci. USA*, **84**, 4398–402.

JOYCE, G. F., VISSER, G. M., VAN BOECKEL, C. A. A., VAN BOOM, J. H., ORGEL, L. E. & VAN WESTRENEN, J. (1984). *Nature*, **310**, 602–4.

KASTING, J. F. (1988). *Icarus*, **74**, 472–94.

KRUGER, K., GRABOWSKI, P. J., ZAUG, A. J., SANDS, J., GOTTSCHLING, D. E. & CECH, T. R. (1982). *Cell*, **31**, 147–57.

LEHMAN, N. & JOYCE, G. F. (1993). *Nature*, **361**, 182–5.

MAHER, K. A. & STEVENSON, D. J. (1988). *Nature*, **331**, 612–14.

MÜLLER, D., PITSCH, S., KITTAKA, A., WAGNER, E., WINTNER, C. E. & ESCHENMOSER, A. (1990). *Helv. Chim. Acta.*, **73**, 1410–68.

SCHIDLOWSKI, M. (1988). *Nature*, **333**, 313–18.

SCHIDLOWSKI, M., HAYES, J. M. & KAPLAN, I. R. (1983). In *Earth's Earliest Biosphere*, ed. J. W. Schopf, pp. 149–86. Princeton, NJ: Princeton University Press.

SCHOPF, J. W. (1993). *Science*, **260**, 640–6.

SCHOPF, J. W. & PACKER, B. M. (1987). *Science*, **237**, 70–3.

SHARPTON, V. L., BURKE, K., CAMARGO-ZANOGUERA, A., HALL, S. A., LEE, S., MARIN, L. E., SUÁREZ-REYNOSO, G., QUEZADA-MUÑETON, J. M., SPUDIS, P. D. & URRUTIA-FUCUGAUCHI, J. (1993). *Science*, **261**, 1564–7.

SHARPTON, V. L., DALRYMPLE, G. B., MARIN, L. E., RYDER, G., SCHURAYTZ, B. C. & URRUTIA-FUCUGAUCHI, J. (1992). *Nature*, **359**, 819–21.

STEVENSON, D. J. (1983). In *Earth's Earliest Biosphere*, ed. Schopf, J. W., pp. 32–40. Princeton, NJ: Princeton University Press.

VON KIEDROWSKI, G. (1986). *Angew. Chem. Int. Ed. Engl.*, **25**, 932–4.

WALTER, M. R. (1983). In *Earth's Earliest Biosphere*, ed. Schopf, J. W., pp. 187–213. Princeton, NJ: Princeton University Press.

ZAHNLE, K. J., KASTING, J. F. & POLLACK, J. B. (1988). *Icarus*, **74**, 62–97.

ZIELINSKI, W. S. & ORGEL, L. E. (1987). *Nature*, **327**, 346–7.

16

The Search for Extraterrestrial Intelligence

ERNST MAYR

How Probable Is It That Life Exists Somewhere Else in the Universe?

What is the chance of success in the search for extraterrestrial intelligence? The answer to this question depends on a series of probabilities. My methodology consists in asking a series of questions which narrow down the probability of success.

Even most skeptics of the SETI project will answer the above question affirmatively. Molecules that are necessary for the origin of life, such as amino acids and nucleic acids, have been identified in cosmic dust, together with other macromolecules, so that it would seem quite conceivable that life could originate elsewhere in the universe. Some of the modern scenarios of the origin of life start out with even simpler molecules, which makes an independent origin of life even more probable. Such an independent origin of life, however, would presumably result in living entities that are drastically different from life on Earth.

Where Can One Expect To Find Such Life?

Obviously only on planets. Even though we have up to now secure knowledge only of the nine planets of our solar system, there is no reason to doubt that in all the galaxies there must be millions if not billions of planets. The exact figure, e.g. for our own Galaxy, can only be guessed.

How Many of These Planets Would Have Been Suitable for the Origin of Life?

There are evidently rather narrow constraints for the possibility of the origin of life on a planet. There has to be a favorable average temperature; the seasonal variation should not be too extreme; the planet must have a suitable distance from its sun; it must have the appropriate mass so that its gravity can hold an atmosphere; this atmosphere must have the right chemical composition to support early life; it must have the necessary consistency to protect the new life against ultraviolet and other harmful radiations; and there must be water on such a planet. In other words, all environmental conditions must be suitable for the origin and maintenance of life.

One of the nine planets of our solar system had the right kind of mixture of these factors. This, surely, was a matter of chance. What fraction of planets in other solar systems will have an equally suitable combination of environmental factors? Would it be 1 in 10, or 1 in 100, or 1 in 1 000 000? It depends on one's optimism which figure one will choose. It is always difficult to extrapolate from a single instance. This figure, however, is of some importance when one is dealing with the limited number of planets that can be reached by the SETI project.

What Percentage of Planets on Which Life Has Originated Will Produce Intelligent Life?

Physicists, on the whole, will answer this question differently from biologists. Physicists still tend to think more deterministically than biologists. They will say: if life has originated somewhere, it will also develop intelligence in due time. The biologist, on the other hand, is impressed by the improbability of such a development. Life originated on Earth about 3.8 billion years ago, but high intelligence did not develop until about half a million years ago. If the Earth had been temporarily cooled down or heated up too much during these 3.8 billion years, intelligence would never have originated.

When answering this question, one must be aware of the fact that evolution never moves on a straight line toward an objective ('intelligence'), as happens during a chemical process or as a result of a law of physics, but that evolutionary pathways are highly complex and resemble more a tree with all of its branches and twigs.

After the origin of life, i.e. 3.8 billion years ago, life on Earth consisted for 2 billion years only of simple prokaryotes, cells without an organized nucleus. These bacteria and their relatives developed surely 50–100 different (some perhaps very different) lineages, but none of them led to intelligence in this enormously long time. Owing to an astonishing, unique event that is even today only partially explained, 1800 million years ago the first eukaryote originated, a creature with a well-organized nucleus and the other characteristics of 'higher' organisms. From the rich world of the protists (consisting of only a single cell) there eventually originated three groups of multicellular organisms: the fungi, the plants and the animals. But none of the millions of species of fungi and plants was able to produce intelligence.

The animals (Metazoa) branched out in the Precambrian and Cambrian to about 60–80 lineages (phyla). Only a single one of them, that of the chordates, led eventually to genuine intelligence. The chordates are an old and well-diversified group, but only one of its numerous lineages, that of the vertebrates, eventually produced intelligence. Among the vertebrates a whole series of groups evolved, types of fishes, amphibians, reptiles, birds, and mammals. Again it was only a single lineage, that of the mammals, that led to high intelligence. The mammals had a long evolutionary history which began in the Triassic, more than 200 million years ago, but only in the latter part of the Tertiary, i.e. some 15–20 million years ago, did higher intelligence originate in one of the *c.* 24 orders of mammals.

The elaboration of the brain of the hominids began only about 3 million years ago, and that of the cortex of *Homo sapiens* only about 300 000 years ago. Nothing demonstrates the improbability of the origin of high intelligence better than the millions of phyletic lineages that failed to achieve it.

How many species have existed since the origin of life? This figure is as much a matter of speculation as the number of planets in our Galaxy. But if there are 30 million living species, and if the life expectancy of a species is about 100 000 years, then one can postulate that there have been billions, perhaps as many as 50 billion species, since the origin of life. Only one of these achieved the kind of intelligence needed for the establishment of a civilization.

To provide exact figures is difficult because the range of variation both in the origination of species and in their life expectancy is so enormous. The widespread, populous species of long geologic duration, usually encountered by the paleontologist, are probably exceptional rather than typical.

How Much Intelligence Is Necessary To Produce a Civilization?

Rudiments of intelligence are found already among birds (ravens, parrots) and among nonhominid mammals (porpoises, monkeys, apes, etc.), but none of these instances of intelligence has been sufficient to found a civilization.

Is Every Civilization Able To Send Signals into Space and To Receive Them?

The answer quite clearly is No. In the last 10 000 years there have been at least 20 civilizations on Earth, from the Indus, the Sumerian and other near Eastern civilizations, to Egypt, Greece and the whole series of European civilizations, to the Mayas, Aztecs and Incas, and to the various Chinese and Indian civilizations. Only one of these reached a level of technology that has enabled it to send signals into space and to receive them.

Would the Sense Organs of Extraterrestrial Beings Be Adapted To Receive Our Electronic Signals?

This is by no means certain. Even on Earth many groups of animals are specialized for olfactory or other chemical stimuli and would not react to electronic signals. Even if there were higher organisms on some planet, it would be rather improbable that they would have developed the same sense organs as we have.

How Long Is a Civilization Able To Receive Signals?

All civilizations have only a short duration. I shall try to emphasize the importance of this point by telling a little fable. Let us assume that there were really intelligent beings on another planet in our Galaxy. A billion years ago their astronomers discovered the Earth and reached the conclusion that this planet might have the proper conditions to produce intelligence. To test this, they sent signals to the Earth for a billion years without ever getting an answer. Finally, in the year 1800 (in our calendar) they decided they would send signals only for another 100 years. When, by the year 1900, no answer had been received, they concluded that surely there was no intelligent life on Earth.

This shows that even if there were thousands of civilizations in the universe, the probability of a successful communication would be extremely slight, owing to the short duration of the 'open window'.

One must not forget that the SETI system is very limited, reaching only part of our Galaxy. The fact that there are a near-infinite number of additional galaxies in the universe is irrelevant as far as the SETI project is concerned.

Conclusions

What conclusions must we draw from these considerations? No fewer than five of the seven conditions to be met for SETI success are highly improbable. When one multiplies these five improbabilities with each other, one reaches an improbability of astronomical dimensions.

Why are there nevertheless still proponents of the SETI project? When one looks at their qualifications, one finds that they are almost exclusively astronomers, physicists and engineers. They are simply unaware of the fact that the success of the SETI project is not a matter of physical laws and engineering capabilities but a matter of biologic and sociologic factors. These, quite obviously, have been entirely left out of the calculations of the possible success of the SETI project.

17

Alone in a Crowded Universe*

JARED DIAMOND

The next time you're outdoors on a clear night and away from city lights, look up at the sky and get a sense of its myriads of stars. Train your binoculars on the Milky Way and appreciate how many more stars escaped your naked eye. Then look at a photograph of the Andromeda nebula as seen through a powerful telescope to realize the enormous number of stars that escaped your binoculars as well. When all those numbers have sunk in, you're ready to ask: How many civilizations of intelligent beings like ourselves must be out there, looking back at us? How long before we are in communication with them, before we visit them or before we are visited?

Many scientists have tried to calculate the odds. Their efforts have spawned a whole new field of science termed exobiology – the sole scientific field whose subject matter has not yet been shown to exist. Since a summary of the calculations fills seven pages of the *Encyclopaedia Britannica*, what more could we learn by further speculation? I'll suggest, nevertheless, that woodpeckers offer a fresh perspective.

Exobiologists find the numbers in their subject matter encouraging. Billions of galaxies each have billions of stars. Many stars probably have one or more planets, and many of those planets probably have an environment suitable for life. Where suitable conditions exist, life will probably evolve eventually.

We ourselves have already launched interplanetary probes. And our progress in freezing and thawing living organisms and in creating life from DNA is leading to techniques that could preserve life for the long duration of an interstellar trip. We are on the brink – surely feasible within a few centuries – of sending out manned interstellar probes. But what about the development of life out there? Less than a few dozen light-years away are several hundred stars, some (most?) of which surely have planets like ours, supporting life. Where are all the flying saucers that we would expect? Where are the intelligent beings that should be visiting us or at least directing radio signals at us?

If intelligent beings from elsewhere had visited Earth after literate civilizations

* Reproduced with permission from *Natural History*, June 1990.

began to develop here several thousand years ago, those beings probably would have searched out the most interesting civilizations, and we would now have written records of the visit. If the visitors had arrived in the preliterate or prehuman past, they might have colonized Earth, and we would have a fossil record of the abrupt arrival of drastically different life forms. Hollywood films depict such visits and tabloids actually claim them. You can see the headlines at any supermarket checkout counter: 'Woman Kidnapped by UFO', 'Flying Saucer Terrorizes Family', and so on. But compare that pseudobombardment, or our expectations, with reality. The silence is deafening.

Let's define an advanced technical civilization as one capable of interstellar radio communication. This is a less demanding definition than flying saucer capability, since our own development suggests that interstellar radio communication will precede interstellar probes. What fraction of the planets in our Galaxy capable of supporting life might support an advanced technical civilization among their life forms? Two arguments suggest that the fraction may be considerable.

First, the sole planet where we are certain that life evolved – our own – did evolve an advanced technical civilization. But – to use the language of statisticians – that argument suffers from the flaws of a small sample size (how can you generalize from one case?) and a high ascertainment bias (we picked out that one case precisely because it evolved our own advanced technical civilization).

The second, and better, argument is that life on Earth is characterized by convergent evolution; that is, many groups of creatures have independently evolved similar physiological adaptations to exploit whatever ecological niche you consider. An example is the independent evolution of flight in birds, bats, pterodactyls and insects. Other spectacular cases are the independent evolution of eyes, and even the devices for electrocuting prey, by many animals. In recent decades, biochemists have recognized convergent evolution at the molecular level, such as the repeated evolution of similar protein-splitting enzymes and membrane-spanning proteins. So common is convergent evolution of anatomy, physiology, biochemistry and behavior that whenever biologists observe some similarity between two species, one of the first questions they ask is: Did that similarity result from common ancestry or from convergence?

If you expose millions of species for millions of years to similar selective forces, of course you can expect similar solutions to emerge time and time again. We know that convergence is very common among species on Earth, and, by the same reasoning, convergence should also exist between Earth's species and those elsewhere. Hence, although radio communication has evolved here only once so far, considerations of convergent evolution lead us to expect its evolution on some other planets as well.

Which brings me at last to woodpeckers. The 'woodpecker niche' is based on digging holes in live wood and on prising off pieces of bark. It's a wonderful niche that offers much more food than do flying saucers or radios. Thus, we might expect convergence among many species that evolved independently to exploit the woodpecker niche. The niche provides dependable food sources in the form of insects living under bark, insects burrowing into wood, and sap. Since wood contains insects and sap year-round, occupants of the woodpecker niche wouldn't have to migrate.

The other advantage of the woodpecker niche is that it is a terrific place for a nest. A hole in a tree is a stable environment with relatively constant temperature and humidity, protected from wind, rain and predators. Other bird species nest in natural holes, but such holes are few in number, quickly become known to predators, get reused year after year and breed infections. Hence, it's a big advantage to be able to excavate a clean, new nest hole in a live tree, instead of having to use a dead tree or natural hole. Other birds pay tribute (unsought by woodpeckers) to that advantage by often usurping woodpeckers' holes.

All these considerations mean that if we're counting on convergent evolution of radio communication, we can surely count on convergent evolution of woodpecking. (For my information about woodpeckers I am indebted to Dr Lester Short, Lamont Curator of Ornithology at the American Museum, whose book *Woodpeckers of the World* is the bible of picologists, scientists devoted to woodpeckers.)

Woodpeckers are very successful birds: there are nearly 200 species, many of them common. They come in all sizes, from tiny birds the size of goldcrests up to crow-sized. They are widespread over most of the world, with a few exceptions that I'll mention later. They don't have to migrate in winter. Some species have even exploited their woodpecking skills to live in treeless places, excavate nest holes in the ground and feed on ants. While the earliest known fossil woodpeckers date only from the Pliocene (about 7 million years ago), molecular evidence indicates that woodpeckers evolved about 50 million years ago.

How hard is it to evolve into a woodpecker? Two considerations seem to suggest, 'Not very hard'. Woodpeckers are not an extremely distinctive old group without close relatives, like egg-laying mammals. Instead, ornithologists agree that the woodpeckers' closest relatives are the honeyguides of Africa, the toucans and barbets of tropical America, and the barbets of the tropical Old World, to which woodpeckers are fairly similar except in their special adaptations for woodpecking. Woodpeckers have numerous such adaptations, but none is remotely as extraordinary as building radios, and all are readily seen as extensions of adaptations possessed by other birds. The adaptations fall into four groups.

First and most obvious are the adaptations for drilling in live wood. These include a strong, straight, chisel-like bill with a hard, horny covering at the tip; nostrils protected with feathers to keep out sawdust; a thick skull; strong head and neck muscles; a broad base of the bill and a hinge between that base and the front of the skull, to help spread the shock of pounding; and possibly a skull design like a bicycle helmet, to protect the brain from shock. These features for drilling in live wood can be traced to features of other birds much more easily than our radios can be traced to any primitive radios of chimpanzees. Many other birds, such as parrots, peck or bite holes in dead wood. Some barbets can excavate in live wood, but they are much slower and clumsier, not as neat as woodpeckers, and peck sideways rather than straight ahead. Within the woodpecker family there is a gradation of drilling ability – from wrynecks, which can't excavate at all, to the many woodpeckers that drill in softer wood, to hardwood specialists such as our sapsuckers and the pileated woodpecker.

Another set of adaptations is for perching vertically on bark, such as a stiff tail to use as a brace; strong muscles for manipulating the tail; short legs; long, curved toes; and a pattern of molting that saves the central pair of tail feathers (crucial in bracing) as the last to be molted. The evolution of these adaptations can be traced even more easily than the adaptations for woodpecking. Even within the woodpecker family, wrynecks and piculets do not have stiff tails for use as braces. But many birds outside the woodpecker family, including creepers and pygmy parrots, do have stiff tails that they evolved to prop themselves on bark.

The third adaptation is an extremely long and extensible tongue, in some woodpeckers fully as long as our own. Once it has broken into the tunnel system of wood-dwelling insects at one point, a woodpecker uses its tongue to lick out many branches of the system without having to drill a new hole for each branch. Some woodpeckers have barbs at the tip of the tongue to spear insects, while others have big salivary glands that make the tongue sticky. Woodpeckers' tongues have many animal precedents, including the similarly long, insect-catching tongues of frogs, anteaters and aardvarks and the brush-like tongues of nectar-drinking lories.

Finally, woodpeckers have tough skins to withstand insect bites plus the stresses from pounding and from strong muscles. Anyone who has skinned and stuffed birds knows that some have much tougher skins than others. Skinners groan when given a pigeon, whose paper-thin skin rips almost as soon as you look at it, but smile when given a woodpecker or hawk.

Thus, while woodpeckers have many adaptations for woodpecking, most of those adaptations have also evolved convergently in other birds or animals, and their unique skull adaptations can at least be traced to precursors. You might therefore expect the whole package of woodpecking to have evolved repeatedly, resulting in many groups of large animals capable of excavating into live wood for food or nest sites. But all the classical evidence, and now the newer molecular evidence, indicates that modern woodpeckers are all more closely related to one another than to any nonwoodpecker. Woodpecking thus appears to have evolved only once.

Picologists take that conclusion for granted. On reflection, though, it's startling to us nonpicologists who had convinced ourselves that woodpecking would evolve repeatedly. Could it be that other pseudowoodpeckers did evolve, but that our surviving woodpeckers were so superior that they exterminated their unrelated competitors? For example, separate groups of mammalian carnivores evolved in South America, Australia and the Old World. But the Old World carnivores (our cats and dogs and weasels) proved so superior that they exterminated South America's carnivorous mammals millions of years ago and are now in the process of exterminating Australia's carnivorous marsupials. Was there a similar shoot-out in the woodpecker niche?

Fortunately, we can test that theory. True woodpeckers don't fly far over water, with the result that they never colonized remote oceanic land masses such as Australia/New Guinea (formerly joined in a single land mass), New Zealand and Madagascar. Similarly, placental terrestrial mammals other than bats and rodents were never able to reach Australia/New Guinea, where instead marsupials evolved good functional equivalents of moles, mice, cats, wolves and anteaters. Evidently, it wasn't so hard to fill those mammalian niches by convergent evolution. Let's see what happened to the woodpecker niche in Australia/New Guinea.

We find there a diverse array of birds that evolved convergently to feed on or under bark, including pygmy parrots, birds of paradise, honeyeaters, Australian creepers, Australian nuthatches, ifritas, ploughbills and flycatchers. Some of those birds have powerful bills used to dig into dead wood. Some of them have evolved elements of the woodpecker anatomical syndrome, such as stiff tails and tough skins. The species that has come the closest to filling the woodpecker niche is not a bird at all but a mammal, the striped possum, that taps on dead wood to detect insect tunnels, rips open the wood with its incisor teeth, and then inserts its long tongue or very long fourth finger to pull out the insects.

But none of these would-be woodpeckers has actually made it into the wood-pecking niche. None can excavate live wood. Many are visibly inefficient: I recall seeing a black-throated honeyeater trying to hop up a tree trunk and repeatedly falling off. The ploughbill and striped possum seem to be the would-bes most effective at digging in dead wood, but both are quite uncommon and evidently can't make a good living by their efforts. New Zealand's and Madagascar's pseudo-woodpeckers are no better. In a stunning instance of convergent evolution, Madagascar's best would-be is also a mammal: a primate called the aye-aye, which operates like a striped possum except for having a very long third, instead of fourth, finger. But just as in Australia/New Guinea, none of the would-bes in New Zealand or Madagascar can excavate in live wood.

Thus, in the absence of woodpeckers, many try and none succeed. The woodpecker niche is flagrantly vacant on those land masses not reached by woodpeckers. If woodpeckers hadn't evolved that one time, a terrific niche would be vacant over the whole Earth, just as it has remained vacant in Australia/New Guinea, New Zealand and Madagascar.

I have dwelt on woodpeckers at length to illustrate that convergence is not universal and that not all opportunities are seized. I could have illustrated the same point with other, equally glaring examples. The most ubiquitous opportunity available to animals is to consume plants, much of whose mass consists of cellulose. Yet no higher animal has managed to evolve a cellulose-digesting enzyme. Those animal herbivores that digest cellulose instead have to rely on microbes housed within their intestines. Among such herbivores, none comes close to achieving the efficiency of ruminants, the cud-chewing mammals exemplified by cows. To take another example, growing your own food would seem to offer obvious advantages for animals, but the only animals to master the trick before the dawn of human agriculture 10 000 years ago were leaf-cutting ants, plus a few other insects that cultivate fungi.

Thus, the evolution of even such obviously valuable adaptations as woodpecking, digesting cellulose efficiently or growing one's own food has proved extraordinarily difficult. Radios do much less for one's food needs and would seem far less likely to evolve. Are our radios a fluke, unlikely to have been duplicated on any other planet?

Consider what biology might have taught us about the inevitability of radio evolution on Earth. If radio building were like woodpecking, some species might have evolved certain elements of the package or evolved them in inefficient form, although only one species managed to evolve the complete package. For instance, we might have found today that turkeys build radio transmitters but no receivers, while kangaroos build receivers but no transmitters. The fossil record might have shown dozens of now-extinct animals experimenting over the last half billion years with metallurgy and increasingly complex electronic circuits, leading to electric toasters in the Triassic, battery-operated rat traps in the Oligocene and, finally, radios in the Holocene. Fossils might have revealed 5 watt transmitters built by trilobites, 200 watt transmitters amid bones of the last dinosaurs, and 500 watt transmitters in use by sabertooths, until humans finally upped the power output enough to broadcast into space.

But none of that happened. Neither fossils nor living animals – not even our closest living relatives, the chimpanzees – had even the most remote precursors of radios. Neither australopithecines nor early *Homo sapiens* developed radios. As recently as 150 years ago, modern *Homo sapiens* didn't even have the concepts that would lead to radios. The first practical experiments didn't begin until about 1888; it's still less than 100 years since Marconi built the first transmitter capable of broadcasting a mere mile; and we still aren't sending signals that could be detected in other galaxies.

I mentioned early in this chapter that the existence of radios on the one planet known to us seemed at first to suggest a high probability of radio evolution. In fact, closer scrutiny demonstrates the reverse. Only one of the billions of species that has existed on Earth showed any proclivities toward radios, and even it failed to do so for the first 69 999/70 000 of its 7 million year history. A visitor from outer space who had come to Earth as recently as A.D. 1800 would have written off any prospects of radios being built here.

You might object that I'm being too stringent in looking for early precursors of radios themselves, when I should instead just look for the two qualities necessary to make radios: intelligence and mechanical dexterity. But the situation there is not encouraging. On the basis of the very recent evolutionary experience of our own species, we arrogantly assume intelligence and dexterity to be the best way of taking over the world. In fact, few animals have bothered with much of either; none has acquired remotely as much of either as we have. Those that have acquired a little of one (smart dolphins, dextrous spiders) have acquired none of the other, and the only other species to acquire a little of both (chimpanzees) has been rather unsuccessful. Earth's really successful species have been dumb and clumsy rats and beetles, which found better routes to their current dominance.

We have still to consider another important variable in calculating the likely number of civilizations capable of interstellar radio communication. That variable is the lifetime of such a civilization. The intelligence and dexterity required to build radios are useful for other purposes that have been the hallmark of our species for much longer than have radios, such as devices for mass killing and means of environmental destruction. We are now so potent at doing both that we are gradually stewing in our civilization's juices. We may not enjoy the luxury of an end by slow stewing. Half a dozen countries have the means to bring us to a quick end, and other countries are eagerly seeking to acquire those means. The wisdom of some past leaders of bomb-possessing nations, or of some present leaders of bomb-seeking nations, does not encourage me to believe that the Earth will have humans and their radios for much longer. Our development of radios was an extremely unlikely fluke; even more of a fluke was our development of them before we developed the technology that will end us in a slow stew or fast bang. While Earth's history of radio civilizations thus offers little hope that they exist elsewhere, its history of the concomitants of radio civilizations suggests that those that might exist anywhere are short-lived.

Thus, the deafening silence from outer space is not surprising. Yes, out there are billions of galaxies with billions of stars. Out there must be some transmitters as well, but not many, and they don't last long. Probably there are no others in our Galaxy. What woodpeckers teach us about flying saucers is that we're unlikely ever to see one. For practical purposes, we're alone in a crowded universe. Thank God!

18

Possible Forms of Life in Environments Very Different from The Earth

ROBERT SHAPIRO AND GERALD FEINBERG

Introduction

We can make only a brief presentation here of our ideas on the extent of life in the universe. A much more detailed account is given in our book *Life Beyond Earth: The Intelligent Earthling's Guide to Life in the Universe* (Feinberg & Shapiro, 1980).

In discussions concerning intelligent life elsewhere, the assumption is often made that such life will develop only in circumstances resembling those on Earth. Estimates are then given of the number of Earth-like planets in our Galaxy, as suitable locations in which life might arise. Such estimates may vary, according to the pessimism or optimism of the observer, from the very few (Hart, 1979) to a billion or more (von Hoerner, 1978). Only those planets are considered habitable which fall into a limited zone around each star. In that zone, liquid water can be present on the surface, and carbon compounds will be abundant. If this view were correct, even in the most optimistic form, then life would be a rare phenomenon, confined to only an insignificant fraction of the material in the universe. From an extreme pessimist's viewpoint, as expressed elsewhere in this book, life may have originated only on the planet Earth. The idea of the specialness of the Earth is of course an old one, and has been expressed many times in theology.

We represent a very different point of view: that the generation of life is an innate property of matter. It can take place in a wide variety of environments very different from the Earth. The life forms that evolve will also be very different from those familiar to us, in harmony with the conditions present there.

Because the point of view opposite to our own is widespread, we think it is important to summarize the reasons presented for it. The most obvious one is the basic unity of Earth life, the only type of life we have encountered. All living things that we are aware of use the same basic set of chemicals to organize their

165

metabolism, with vital roles for proteins and nucleic acids. A person in a Chinese village might similarly be convinced that only the Chinese language existed, as it was the only one spoken by all the people he knew. In fact, no firm conclusion can be drawn at all from cases where only a single example of a phenomenon is at hand, and a good strategy would be to search for additional examples.

Other arguments have been made on the basis of scientific principles, and we have found that their adherents could be grouped into two categories, which we have called 'predestinist' and 'carbaquist'.

The Predestinists

The predestinists hold that the limited set of chemicals that characterize our biochemistry will be the inevitable result of random chemical synthesis, throughout the universe. We can express this in direct quotes: 'The implications of these results is that organic syntheses in the Universe have a direction that favors the production of amino acids, purines, pyrimidines and sugars: the building blocks of proteins and nucleic acids. Taken in conjunction with the cosmic abundance of the light elements, this suggests that life everywhere will be based not only on carbon chemistry, but on carbon chemistry similar to (although not necessarily identical with) our own.' (Horowitz, 1976) 'The essential building blocks of life – amino acids, nitrogen-bearing heterocycle compounds and polysaccharides – are formed in space. These compounds occur in large quantities throughout the galaxy.' (Hoyle & Wickramasinghe, 1977)

These concepts are imaginative but unfortunately not supported by the facts. The presence of organic compounds in interstellar dust clouds has been demonstrated by spectroscopy. (Mann & Williams, 1980) However, the molecules that have been identified do not include the essential building blocks of Earth life, such as amino acids, sugars, fatty acids and the bases of nucleic acids. Small molecules, fragments of molecules and exotic substances such as cyanopolyynes (Freeman & Millar, 1983) and ionized polycyclic aromatic hydrocarbons (Szczepanski & Vala, 1993) are featured instead. Meteorite analyses provide another source of information about extraterrestrial organic chemistry (Cronin et al., 1988). Amino acids are present there, but as part of a complex mixture containing many compounds irrelevant to our biochemistry. There is no preference for the synthesis of compounds important to Earth life. Finally, we can consider prebiotic simulation experiments of the kind initiated by Stanley Miller and Harold Urey (1959). Good yields of amino acids have frequently been obtained. Although there is considerable doubt that the conditions used do represent those of the early Earth (Kasting, 1993), such experiments do demonstrate that the preparation of amino acids from very simple compounds is feasible. They do not, however, show that the result is inevitable. In fact, the very first effort by Stanley Miller (1974) was unsuccessful: 'The next morning there was a thin layer of hydrocarbon on the surface of the water, and after several days, the hydrocarbon layer was somewhat thicker.' No amino acids were detected at all.

The course of organic synthesis in the universe may on some occasions turn in the direction of the chemicals of Earth life, but it is clear that it can take other directions as well. We are not the inevitable predestined endpoint of cosmic evolution.

The Carbaquists

A somewhat different point of view is taken by the carbaquists. They feel that life elsewhere in the universe must resemble that present on Earth, because no other basis for life can exist. Our own chemical system, particularly in its use of water, as a solvent, and carbon, as the key building block of large molecules, is uniquely fit for the purpose of sustaining life. The notion of the fitness of our environment was advocated early in this century by the American chemist Lawrence Henderson (1913) and has had a number of advocates more recently. Again, we shall quote them directly: 'I have become convinced that life everywhere must be based primarily upon carbon, hydrogen, nitrogen and oxygen, upon an organic chemistry therefore much as on the Earth, and that it can arise only in an environment rich in water' (Wald, 1974). '. . . so I tell my students: learn your biochemistry here and you will be able to pass examinations on Arcturus' (Wald, 1973). 'The capacity for generating, storing, replicating and utilizing large amounts of information implies an underlying molecular complexity that is known only among compounds of carbon' (Horowitz, 1976).

Water *does* have certain special properties as a solvent. One is the greater amount of heat that is needed to melt, warm up or boil a quantity of water, as compared with the heat needed for most other solvents. Bodies of water thus tend to stabilize the climate of their environment. This property may be pleasant, but it is hardly essential to life. Other physical features could work to stabilize a climate. Alternatively, living beings could adapt by many strategies, such as spore formation or migration, to changes in temperature. Another property of water that is greatly admired by the carbaquists is its polar character. Water is a good solvent for charged substances, and an appropriate medium for the conduct of transformations of such substances. However, an enormous number of reactions have been described by chemists which take place in less polar, or nonpolar, solvents. The absence of water is in some cases essential to the course of the reaction. There is no reason why such reactions could not be the basis of a metabolic scheme to sustain life.

The special properties of carbon include its ability to bond to itself in long chains and to form bonds to four other atoms at one time. An enormous number of compounds containing carbon can, therefore, exist. It is not necessary for an atom to bond to itself to form long chains, however: the chains could be made of two or more atoms in alternation. Furthermore, it is conceivable that a basis for life could be constructed using an alternative chemistry in which the possibilities were not as vast as those of carbon. The situation can be compared with the problem of information storage in printed form. The English language uses 26 capital letters, 26 small letters, space and a number of punctuation marks to do this: about 60 characters in all. The same amount of information can be stored by a computer, using only the two symbols 1 and 0. Six lines must be used to hold the contents of one English line, but the data are stored just the same. In the same way, a less complex chemistry could serve as the genetic basis for life, perhaps with a larger number of components needed in each molecule, cell or other unit.

The carbaquist viewpoint cannot be made convincing by the type of reasoning its adherents have presented, but also it cannot be refuted strictly on the basis of reason and analogy. We can best proceed by searching for alternative life forms. The discovery of one, with a different physical or chemical basis, would quickly settle the issue. Failure to do so, after a number of extraterrestrial life forms had been encountered, would move us toward acceptance of the carbaquist argument.

A Definition and Some Conditions

In seeking a broader view of the possibilities of life in the universe than that put forward by the carbaquists and predestinists, we have started by framing a definition of life that is independent of the local characteristics of Earth life: Life is the activity of a biosphere. A biosphere is a highly ordered system of matter and energy characterized by complex cycles that maintain or gradually increase the order of the system through an exchange of energy with its environment.

An important feature in our definition is the identification of the biosphere as the unit of life. The history of life on Earth then becomes the tale of the continuous survival and evolution of the biosphere from its origin on the prebiotic Earth. Replication, and subdivision into organisms and species have been strategies adopted by our own biosphere to ensure its own survival but they need not be the methods used by an extraterrestrial biosphere.

In searching for extraterrestrial life, apart from the special case of intelligent life, we should look for biospheres, rather than looking specifically for individual living things. Until 60 years ago, the best remote indicator of the presence of life on Earth did not arise from human actions but from the presence of oxygen in the atmosphere, due to the activity of blue-green algae and denitrifying bacteria. The same principle may apply as well in other cases. The specific ways in which we seek other biospheres will depend on the types of biosphere that we are seeking (Feinberg & Shapiro, 1980).

The association of life with increasing order, and the emphasis on the need of a flow of energy to sustain this, are related to the ideas of the physicists Erwin Schrödinger (1956) and Harold Morowitz (1968). Our biosphere has a number of easily recognizable forms of order, all in a very high degree. It contains very large numbers of a few thousand small organic molecules and none (or very few) of millions of other molecules. The presence of only a few distinct types of large molecules in living things is a second type of order. An additional dramatic type is the near identity of the base sequences in the DNA in different cells in a multicelled organism. If comparable amounts of order (though perhaps of a very different type) were to be found in an extraterrestrial environment, this would be a strong sign of the presence of life.

With the definition in hand, we can now make a list of several conditions that must be met if life is to originate and develop in a particular location.

1. A flow of free energy. Any of a number of different types could suffice, such as light energy and chemical energy, as on Earth, but also other forms of electromagnetic radiation, such as infrared light and X-rays. Other forms of energy such as streams of charged particles, heat differentials and nuclear energy could also be used.

2. A system of matter capable of interacting with the energy and using it to become ordered. The nucleic acids and proteins of Earth need not be the basis of this order. In fact, it need not depend on chemical reactions at all but could be based on physical processes such as the movement of particles, or molecular rotations. However, some systems will be superior to others. A liquid or a dense gas is preferable to a solid for the conduct of reactions. Helium is a poor choice as the base for the development of an ordered system based on a multiplicity of chemical compounds.

3. Enough time to build up the complexity that we associate with life. This is a critical question that determines the scope of life in the universe. The rate of a process, such as the chemical reactions between molecules widely scattered in outer space, may be so slow that the entire lifetime of the universe to date has been insufficient for appreciable order to develop. In another case, a drastic change in an environment, e.g. the conversion of a star to a supernova, could destroy the base upon which order has been accumulating.

Alternative Bases for Chemical Life

With the requirements in mind, we have tried to generate some speculative suggestions for life forms that would function on a chemical basis different from that on Earth.

1. Life in ammonia. Liquid ammonia, perhaps with some water, would serve as the solvent. This could occur on a planet with temperatures near -50 °C. Weaker chemical bonds, such as nitrogen, would predominate in metabolism.
2. Life in hydrocarbons. A mixture of hydrocarbons would function as the solvent. A wide range of temperatures could be accommodated, depending on the composition of the mixture. Processes involving charged species would play only a small role. Reductive reactions, such as hydrogenation, could be used as an energy source.
3. Silicate life. Silicates exhibit a rich chemistry with a propensity for forming chains, rings and sheets, as in minerals on Earth. Cairns-Smith (1982) has argued that layers of crystalline silicates functioned as a primitive life form on the early Earth, and gradually changed, through evolution, into carbon-based life forms. Another basis for silicate life could also occur in much hotter environments. At a temperature above 1000 °C, the medium would become liquid, and could serve as a basis for evolving chemical order. This process could occur on a planet quite close to its sun, or in the molten interior of a planet, such as Earth.

Alternative Chemical Life within Our Solar System

Locations within our own solar system offer the best opportunities for the detection of alternative life forms in the immediate future. Some of the more promising possibilities are listed below:

1. Earth: in interior magma, or within specialized niches on the surface.
2. Mars: if life is present, it is probably based on carbon and water.
3. Jupiter: many possibilities exist in varied environments.
4. Europa: life within water oceans beneath the crust.
5. Io, Venus: life in liquid sulfur.
6. Titan: ammonia- or hydrocarbon-based life.

Physical Life

As we suggested earlier, many physical processes may serve for the storage and increase of order. We have called such systems 'physical life', and list some speculative possibilities below.

1. Plasma life within stars. Such life would be based upon the reciprocal influence of patterns of magnetic force and the ordered motion of charged particles. It could exist within our own Sun. Individual creatures are called 'plasmobes'.
2. Life in solid hydrogen. This could occur on a planet with a temperature of only a few Kelvin. Infrared energy would be absorbed and stored in the special arrangement of *ortho-* and *para*-hydrogen molecules.
3. Radiant life. Life would be based upon the ordered patterns of radiation emitted by isolated atoms and molecules in a dense interstellar cloud. Such clouds can have a long lifetime before they collapse. At a density of 10^4 atoms per cubic centimeter, they may last for millions of years. Individual beings, called 'radiobes' could exist.
4. Life in neutron stars. Such life would be based on the properties of the polymeric atoms whose existence was proposed by physicist Malcolm Ruderman (1974). It is possible that such polymer chains could store and transmit information in a way that bears an eerie similarity to the functions of nucleic acids.

It may be difficult to think of such systems as being alive, with the capability in some cases of developing organisms, complex ecologies and even civilization. We must remember that the association of a protein with a nucleic acid, when viewed abstractly, also does little to convey the wonders, such as elephants and *Sequoia* trees, that ultimately arise from it.

Conclusions

Our examples have not been presented in order to make specific predictions, but rather to suggest the vast variety of forms that life elsewhere may take. If we were gifted with a vision of the whole universe of life, we would not see it as a desert, sparsely populated with identical plants which can survive only in rare specialized niches, but rather as a botanical garden with countless species, each thriving in its own setting. This garden awaits our exploration.

References

CAIRNS-SMITH, A. G. (1982). *Genetic Takeover and the Origin of Life*. Cambridge: Cambridge University Press.

CRONIN, J. R., PIZZARELLO, S. & CRUIKSHANK, D. P. (1988). In *Meteorites and the Early Solar System*, ed. J. F. Kerridge and M. S. Matthews, p. 818. Tucson, AZ: University of Arizona Press.

FEINBERG, G. & SHAPIRO, R. (1980). *Life Beyond Earth. The Intelligent Earthling's Guide to Life in the Universe*. New York: William Morrow.

FREEMAN, A. & MILLAR, T. J. (1983). *Nature*, **301**, 402.

HART, M. H. (1979). *Icarus*, **37**, 351.

HENDERSON, L. J. (1913). *The Fitness of the Environment*, New York: Macmillan.

HOROWITZ, N. H. (1976). *Accounts Chem. Res.*, **9**, 1.

HOYLE, F. & WICKRAMASINGHE (1977). *New Scientist* (17 Nov.), 174.

KASTING, J. F. (1993). *Science*, **259**, 920.

MANN, A. P. C. & WILLIAMS, D. A. (1980). *Nature*, **283**, 721.

MILLER, S. L. (1974). In *The Heritage of Copernicus: Theories Pleasing to the Mind*, ed. J. Neyman p. 228. Cambridge, MA: MIT Press.

MILLER, S. L. & UREY, H. C (1959). *Science*, **130**, 245.

MOROWITZ, H. (1968). *Energy Flow in Biology*. New York: Academic Press.

RUDERMAN, M. (1974). In *Physics of Dense Matter*, ed. C. Hansen. Dordrecht: Reidel.

SCHRÖDINGER, E. (1956). *What Is Life?*, New York: Anchor Books.

SZCZEPANSKI, J. & VALA, M. (1993). *Nature*, **363**, 699.

VON HOERNER, S. (1978). *Naturwissenschaften*, **65**, 553.

WALD, G. (1973). In *Life Beyond Earth and the Mind of Man*, ed. R. Berendzen, p. 15. Washington, DC: NASA Scientific and Technical Information Office.

WALD, G. (1974). In *Cosmological Evolution and the Origins of Life*, ed. J. Oró *et al.*, vol. I, p. 7. Dordrecht: Reidel.

19

Cosmological SETI Frequency Standards

J. RICHARD GOTT, III

Introduction

In 1973 Drake and Sagan proposed a SETI frequency standard of $\nu_0 \sim 56$ GHz tied to the observed cosmic microwave background $h\nu_0 = kT_0$, where T_0 is the current temperature of the cosmic microwave background. They noted that a transmitting civilization in a distant galaxy will, however, have transmitted its signal to the Earth at an earlier cosmological epoch when T was larger than is measured today, tending to increase the 'natural' frequency, but that the cosmological Doppler effect will tend to decrease the frequency. Not knowing of their work, I proposed this same frequency standard ($h\nu_0 = kT_0$: Gott, 1982) in the first edition of this book. I had noticed that the two effects mentioned above in fact cancel each other out exactly (which was a new result), so that this frequency standard was indeed universal. If a transmitting civilization is at a redshift z, it will observe a microwave background temperature of $T_1 = T_0 (1 + z)$ and will emit signals at a frequency of $h\nu_e = kT_1 = kT_0 (1 + z)$, but because of the cosmological Doppler shift we will observe these transmitted photons at a frequency $h\nu_0 = h\nu_e (1 + z)^{-1}$ so that we observe $h\nu_0 = kT_0 (1 + z) (1 + z)^{-1} = kT_0 \sim 56$ GHz independent of the redshift of the emitting civilization. This is very important because it means that even observers in distant galaxies can use a universal frequency standard to communicate. Also, it is quite important that the frequency standard $h\nu_0 = kT_0$ was discovered independently by two groups (Drake & Sagan, 1973, and, Gott, 1982). The frequency standard must be simple enough so that different people (or civilizations) can think of it independently. Interestingly, the 21 cm line for SETI was also thought of independently by two groups: it was discovered by Cocconi and Morrison (1959) and at a similar epoch was independently thought of by Drake for project Ozma (cf. Drake & Sobel, 1992, for discussion). In my paper I also suggested for ground-based observations the frequency standard $h\nu = kT_0/2$, but Drake and Sagan did not, so it has not yet been discovered independently more than once. In my paper I stressed that the frequency standard would be established with much higher accuracy after the COBE satellite had done its work measuring the temperature and isotropy of the cosmic microwave background. This has now occurred. The COBE, FIRAS experiment has now measured the temperature of the cosmic microwave background to be $T_0 = 2.726$ K $\pm$ 0.01 K (2σ) (Mather *et al.* 1993). This represents an improvement in accuracy by about a factor of 27 since I wrote my paper. Using this latest COBE value of the temperature, we find the cosmological SETI frequency to be

$$\nu_0 = 56.8 \text{ GHz} \pm 0.2 \text{ GHz} \ (2\sigma).$$

The Number of Habitable Planets

To discuss the possibilities for extragalactic communication we would like to calculate the number of habitable planets in the observable universe, and to do so we must determine the local density of habitable planets and use the appropriate cosmological model.

The discovery of the 2.7 K cosmic microwave background by Penzias & Wilson (1965) has led to virtually unanimous acceptance of the big bang model for the origin of the universe. The simplest big bang models are the Friedmann models with zero cosmological constant which represent solutions to Einstein's field equations. These models are completely characterized by their values of H_0 and Ω. Hubble's constant H_0 measures the current expansion rate of the universe. Current estimates of H_0 fall in the range 50–100 km s^{-1} Mpc^{-1}. For convenience throughout this chapter I shall adopt $H_0 = 50$ km s^{-1} Mpc^{-1} (cf. Sandage & Tamman, 1975). The quantity $\Omega = 8\pi G\rho/3H_0^2$ measures the ratio between the current density and the critical density required to eventually halt the expansion of the universe. If $\Omega > 1$, the universe has positive curvature, is closed with finite volume, and, at some point in the future, the expansion of the universe will halt and the universe will contract again to a dense state. If $\Omega = 1$, the universe is spatially flat (euclidean) with infinite volume and will expand forever. If $\Omega < 1$, the universe is negatively curved, has infinite volume, and will expand forever. I shall consider all three types of cosmology.

The number of habitable (Earth-like) planets in our Galaxy is estimated to be of order 10^9 (Bracewell, 1975). The total blue luminosity of our Galaxy is of order $L = 1.5 \times 10^{10} L_{Sun}$. Gott & Turner (1976) estimate the mean blue luminosity density of the universe to be $\rho_L \sim 4.7 \times 10^7 L_{Sun}$ Mpc^{-3}. Combining these estimates gives a mean cosmological density for habitable planets of $\rho_{HP} = 3 \times 10^6$ (HP) Mpc^{-3}.

Consider a cosmology with $\Omega > 1$. This universe has the geometry of a three-sphere. As an example take $\Omega = 2$. Then the radius of curvature of the universe is $a_0 = cH_0^{-1} = 6000$ Mpc and it currently has a total finite volume of $V = 2\pi^2 a_0^3 = 4.3 \times 10^{12}$ Mpc^3. The number of habitable planets in the universe is $\rho_{HP} V = 10^{19}$. There will be other intelligent species in the universe so long as the probability for a habitable planet to spontaneously develop intelligent life is greater than 10^{-19}.

If $\Omega \leqslant 1$, the universe has an infinite number of habitable planets, and as long as there is a finite probability of forming intelligent life on each habitable planet, there will be an infinite number of intelligent civilizations in the universe. If the probability is very small, however, the nearest of these civilizations will be very far away.

The above models are those with the simplest topologies. Einstein's field equations tell us about the geometry of spacetime but not its topology. For example, a space-like slice of the $\Omega = 1$ cosmology is an infinite euclidean three-dimensional space. One can construct a cube with side L in this space and topologically identify the top with the bottom, the front with the back and the left with the right side. In this way one can make an $\Omega = 1$ universe with a T^3 toroidal topology and a finite volume $V = L^3$. Lower limits on L can be set by our failure to find multiple images of famous superclusters of galaxies. Gott (1980) finds L must be >400 Mpc. Thus, if $\Omega = 1$, there must be at least $\rho_{HP} V = 2 \times 10^{14}$ habitable planets in the universe (and an infinite number if the topology is simple). There are also multiply connected $\Omega < 1$ cosmologies. These all have $V > 0.98$ a_0^3 where $a_0 \geqslant 6000$ Mpc is the radius of curvature of the negatively curved space (cf. Gott, 1980). Thus, the $\Omega < 1$ cosmologies must have at least 6×10^{17} habitable planets (and an infinite number if the topology is simple).

Because of the finite age of the universe, we cannot see all of it by the present epoch. The current limit of observation is called the particle horizon. At the present time we can see everything within the horizon radius (r_H). Photons from beyond r_H have simply not had time to reach us by the present epoch. Big bang models have ages that are comparable with H_0^{-1} and so a crude estimate of the number of habitable planets within the horizon is $4\pi\rho_{HP} r_H^3 / 3 = 3 \times 10^{18}$, where $r_H \sim cH_0^{-1} = 6000$ Mpc. To compute an exact estimate, one must use the curved spacetime metrics of the appropriate cosmological model. I have performed this calculation for three representative cases using the standard Friedmann metrics. The numbers of habitable planets within the horizon are 6×10^{18} for $\Omega = 2$, 2×10^{19} for $\Omega = 1$, and 2×10^{21} for $\Omega = 0.1$. These numbers are not many orders of magnitude away from the crude estimate. Even if the universe is infinite, we only see a finite portion of it at the present epoch. If the probability of forming intelligent life on a habitable planet is less than 5×10^{-22}, then we are unique within the currently observable universe.

As times goes on, the horizon radius increases and we see more and more of the universe. Dyson (1979) has shown that in an $\Omega \leq 1$ cosmology it would not violate thermodynamics for a supercivilization to maintain itself indefinitely. Conserving energy by operating at lower and lower temperatures as the universe expands and by interspersing periods of hibernation, the supercivilization could, in principle, produce an infinite number of thoughts in an infinite time if $\Omega \leq 1$ (although insurmountable practical problems may be encountered). The spectrum of fluctuations seen by the COBE satellite is, within the observational errors, consistent with the Zeldovich (1972) spectrum predicted by modern inflationary (Guth, 1981) models. In these models, the current value of Ω is typically very close to 1 and the size of our inflationary domain which will eventually come within the particle horizon in the far future is very large. As an example consider the Linde (1990) chaotic inflationary model in which the universe will continue its normal expansion with $\Omega \sim 1$ for the next $10^{5 \times 10^6}$ years. At that time the number of habitable planets that will have come within the horizon will be larger than the current value of 2×10^{19} by a factor of $\sim 10^{5 \times 10^6}$. Thus, even if the probability of forming an intelligent civilization on a habitable planet is extremely low (say $P_i = 10^{-3000}$: Hart, 1982), then we still expect of order $10^{5 \times 10^6}$ intelligent civilizations in our inflationary domain (which will eventually become visible).

Cosmic Frequency Standards

For communication within the Galaxy, the 21 cm line proposed by Cocconi & Morrison (1959) is promising. All the observers can agree on a velocity standard (such as the velocity of the center of the Galaxy) and tune their receivers to that standard. Alternatively, signals can be sent to specific target stars so that the signal will have the 21 cm line frequency in the rest frame of the target star. Several unsuccessful 21 cm searches have been undertaken so far. The data in the previous section would indicate that it is profitable to look for life beyond our own galaxy. To do this we must find cosmological frequency standards for communication.

Frequency standards based on line radiation such as the 21 cm line are not useful for intergalactic communication because of the cosmological redshift. In principle, one could identify the 21 cm rest wavelength in each galaxy to be studied (redshift could be measured to an accuracy of ~ 10 km s^{-1}, giving a frequency standard of $\Delta v/v = 3 \times 10^{-5}$). But this would be very time-consuming for a survey out to 600 Mpc.

Fortunately, the cosmic blackbody radiation discovered by Penzias & Wilson (1965) provides a frequency standard with the desired properties. It has a Planck spectrum:

$$I(v)dv = \frac{2\pi h v^3}{c^2} \frac{1}{\exp(hv/kT) - 1} dv \qquad (19.1)$$

with a current measured characteristic temperature $T = 2.726$ K. This spectrum is consistent with the observations over the range 400 MHz $< v <$ 800 GHz (cf. Weiss, 1980; Mather et al., 1993). As the universe expands by a factor (a_2/a_1) from time t_1 to t_2, each photon is redshifted by the expansion factor; thus $v_2 = (a_1/a_2)v_1$. The blackbody radiation spectrum retains the form of eq. (19.1) with a value of the temperature at time t_2 of $T_2 = (a_1/a_2)T_1$.

Let a civilization measure the cosmic microwave background and determine its temperature, T_e. Then let that civilization emit a signal with frequency v_e given by $hv_e = kT_e$. When the photon is received by an observer, its frequency is measured to be $v_r = (a_e/a_r)v_e$ and at that epoch the observer will see a blackbody radiation spectrum with $T_r = (a_e/a_r)T_e$. Thus, the observer finds $hv_r = kT_r$. Once a photon is emitted with $hv = kT$, it will retain that value always as the universe expands, since the other photons in the blackbody radiation redshift in exactly the same fashion as it does. Civilizations can send signals with $hv = kT$ knowing that they can be received *at any time in the future*, potentially reaching a great number of galaxies. Observers can search the entire sky at $hv = kT$ and detect signals from galaxies regardless of redshift. With $T_0 = 2.726$ K, $v_0 = 56.8$ GHz. The Earth has a velocity of approximately $V_p = 300$ km s^{-1} with respect to the rest frame established by the cosmic blackbody radiation. This produces a fractional variation $\delta T/T \sim V_p/c \sim 10^{-3}$ as a function of direction in the sky. However, once the local motion has been determined from the blackbody data, it can be taken out and a rest standard relative to the microwave background established. If one wishes to observe in a direction X, one uses $v_0 = kT(X)/h$ where $T(X)$ is the value of the background temperature in direction X. For sending signals in direction X one uses $v_0 = kT(-X)/h$ where $-X$ is the direction antipodal to X. This technique should reduce the uncertainty in v_0 to less than one part in 10^5, the observed level of cosmological fluctuations on small and intermediate angular scales (10°) as found by the COBE satellite (Smoot et al., 1992). Fluctuations of similar magnitude can be produced by scattering of the microwave photons in gas in clusters of galaxies. In the future we should be able to measure the absolute temperature of the microwave background over the sky to sufficient accuracy to allow us to establish v_0 to one part in 10^5 in any direction. This would require us to search only a bandwidth of $\pm \delta v \sim 10^{-5}v_0 \sim 568$ kHz. With only our present knowledge of the absolute temperature $T = 2.726 \pm 0.01$ K (95% confidence limits: Mather et al., 1993), we would have to search a band of $\pm$ 208 MHz about 56.8 GHz.

The simplest cosmic frequency standard would be $hv_0 = kT$. In these units the blackbody spectrum has a particularly simple form: at low frequencies $I(v)dv \propto v^2$ (Rayleigh–Jeans law), v_0 is the frequency at which $I(v)dv$ falls a factor of $(e-1)$ below the Rayleigh–Jeans law. We expect civilizations to know of the pure number e (the base of the natural logarithms) and for kT to be a natural unit of energy. A gas in thermal equilibrium with the blackbody radiation will have a mean energy per particle of $kT/2$ per degree of freedom. A monotonic gas will have a mean energy per particle of $3kT/2$. Thus the frequencies $v_1 = 28.4$ GHz $= kT/2h$, $v_0 = 56.8$ GHz $= kT/h$, $v_2 = 85.2$ GHz $= 3kT/2h$ naturally suggest themselves. The maximum intensity of the microwave background in frequency units $I(v)_{max}$ occurs at $hv_{max} = 2.81214kT$ (159.7 GHz). The maximum emission in wavelength units occurs at $hv_3 = 4.9651kT$ (282.0 GHz). The maximum photon emission in wavelength units occurs at $hv_4 = 3.9207\,kT$ (222.7 GHz). The mean energy per photon in the blackbody spectrum is $v_5 = 2.7011\,kT$ (153.4 GHz). (The frequency values above assume $T = 2.726$ K, since T is currently known (95% confidence level) to an accuracy of about ± 1 part in 300; the frequencies are also currently known to the same precision. Ultimately, the frequency standards could be established to an accuracy of one part in 10^5.) With suitable imagination, perhaps a dozen plausible transmission frequencies associated with the microwave background could be found.

Experiments

The minimum detectable flux P_r (W m^{-2}) with a required signal-to-noise ratio of S/N is given by

$$P_r = \frac{kT_s}{A_e} \frac{\Delta v}{\sqrt{\Delta v \tau}} (S/N) \qquad (19.2)$$

where $A_e = \eta(\pi r^2)$ is the effective area of an antenna of radius r, Δv is the bandwidth of a channel and τ is the detection time. $\eta = 0.9$ is the efficiency of the antenna. T_s is the system noise temperature, the sum of the background noise temperature T_N plus that due to amplifier noise, antenna imperfections and atmospheric noise. We assume that the signal has an intrinsic bandwidth that is less than our channel bandwidth. In space, for $v > 3$ GHz the value of T_N is

$$kT_N \simeq \frac{hv}{\exp(hv/kT) - 1} + hv \qquad (19.3)$$

The first term is due to the microwave background radiation and the second is due to quantum noise (cf. Oliver, 1977). For $hv < kT$, $T_N \approx T = 2.726$ K, for $hv > kT$, $T_N \sim hv/k > 2.726$ K. Thus, at frequencies significantly higher than

Table 19.1. *Possible experiments*

	I	II
Frequency	56.8 GHz	28.4 GHz
searched	±568 kHz	±104 MHz
Antenna diameter	2.4 m	0.3 m
Location	Space	Ground
Bandwidth	0.02 Hz	0.04 Hz
No. of channels	28×10^6	28×10^6
τ	50 s	25 s
T_{sys}	10 K	100 K
All-sky search	10.5 yr	1.9 yr
Minimum detectable		
flux with $S/N = 10$	8×10^{-24} W m^{-2}	8×10^{-21} W m^{-2}

$v_0 = 56.8$ GHz, detection becomes appreciably more difficult because of the quantum noise: $T_N (v_1) = 3.5$ K, $T_N (v_0) = 4.3$ K, $T_N (v_{max}) = 8.2$ K. This makes the frequencies $v_1 = 28.4$ GHz and $v_0 = 56.8$ GHz preferable to v_{max}, v_2, v_3, v_4, v_5.

Because of a wide O_2 absorption band at 60 GHz, a search at $v_0 = 56.8$ GHz must be carried out from space, while 28.4 GHz is observable from the ground. Table 19.1 lists parameters for two possible experiments of varying expense. These experiments could reach interesting flux limits comparable with those in other SETI searches (cf. Horowitz and Sagan, 1993). Experiment I represents what ultimately could be done with current technology once the absolute temperature of the microwave background is known (by future COBE-type satellites) to an accuracy of 1 part in 10^5. Experiment II could be done with current knowledge of the microwave background at moderate expense from the ground. The COBE satellite DMR experiment has two receivers covering the range 53 GHz ± 0.5 GHz (the others are at 31.5 GHz and 90 GHz), which is unfortunately not at the SETI frequency. Now the FIRAS experiment which measures the entire spectrum has observed at 56.8 GHz, but so far the only announced FIRAS results have extended from 60 GHz to 600 GHz. Results below 60 GHz have not been announced because FIRAS is more difficult to calibrate at lower frequencies (Ned Wright, private communication). But the FIRAS data could be searched for point sources at 56.8 GHz. Unfortunately, since FIRAS is essentially a one-channel receiver, it has relatively poor sensitivity compared with the proposed multichannel experiments described in table 19.1. The FIRAS experiment has a sensitivity of

10^{-9} W m^{-2} sr^{-1} (C. Bennett, private communication) and a 7° full-width-half-maximum diameter beam with sky coverage of $\sim$0.01 sr, so for a point source it would have the sensitivity to detect a flux of 10^{-11} W m^{-2} which is considerably less sensitive than the experiments described in table 19.1.

Conclusions

In this chapter, the possibility of detecting civilizations at cosmological distances is examined. The first step is to calculate the number of habitable planets. If there are of order 10^9 habitable planets in our Galaxy, the mean cosmological density of habitable planets is of order 3×10^6 HP Mpc^{-3} (with $H_0 = 50$ km s^{-1} Mpc^{-1}). If the universe is simply connected, then the number of habitable planets in the universe is $>10^{19}$ if $1 < \Omega < 2$ and is infinite if $\Omega \leq 1$. Let P_i be the probability of an intelligent civilization arising spontaneously on a habitable planet. If the universe is simply connected, we are not unique if P_i is greater than 10^{-19}. Even if there are currently no other civilizations visible in our Galaxy, there may be many civilizations visible in external galaxies, and if there are other civilizations visible in our Galaxy, the number of extragalactic civilizations visible within 2 billion light-years is expected to be 7×10^5 times larger than the number of civilizations visible in our Galaxy.

So, even if intelligent life forms frequently, there are still many more opportunities to detect life at large distances. In this case the relative detectability of civilizations at cosmological distances depends on the luminosity function of radio transmitters. If it is sufficiently flat, there is even a possibility that the brightest transmitter observed is extragalactic. Dyson (1960) has described a process by which an advanced civilization can utilize the entire energy of its star. Kardashev (1964) classified civilizations by their power output: Type I $= 4 \times 10^{12}$ W, Type II $= 4 \times 10^{26}$ W $= 1\ L_{Sun}$, Type III $= 4 \times 10^{37}$ W $= 10^{11}L_{Sun}$. If civilizations utilize 10^{-4} of their energy resources for transmitting, then either experiment I or II (table 19.1) could detect a Kardashev Type III civilization anywhere in the observable universe (i.e. within the current particle horizon). Experiment II could detect a Type II civilization out to a distance of 67 000 light-years, while Experiment I could detect a Type II civilization out to a distance of 2.1 million light-years.

For communication at cosmological distances, line radiation such as 21 cm is not very useful because of the cosmological redshift. However, we can establish frequency standards tied to the $T_0 = 2.726$ K cosmic microwave background discovered by Penzias & Wilson (1965). Two promising frequencies are 56.8 GHz ($h\nu_0 = kT_0$) and 28.4 GHz ($h\nu = kT_0/2$). With careful measurements of the microwave background a civilization could set the frequency standard to an accuracy of one part in 10^5. A space experiment is required to observe 56.8 GHz, but modest searches at 28.4 GHz can be conducted from the ground with present technology, table 19.1.

Recently I have analyzed SETI prospects using the Copernican Principle (Gott, 1993). I argue that out of all the places for intelligent observers to be, there are only a few special places and many nonspecial places, so that you are simply likely to be at one of the many nonspecial places. With this hypothesis one can make many predictions. It is interesting to note that Hart's (1975) argument that we must be alone in the Galaxy – otherwise we would have been colonized by extraterrestrials – also relies implicitly on the Copernican Principle by arguing that (*assuming* colonization occurs not infrequently) if there were many intelligent civilizations to form in our Galaxy, it would be improbable for us to be the first. Hart's (1975) paper is based on Fact A: we have not been previously colonized by extraterrestrials. Part of my paper (Gott, 1993) is about what I would call Fact B: we have thus far not colonized the Galaxy. As I argue there, Fact B together with the Copernican Principle is sufficient *by itself* to explain Fact A. Since we observe Fact B, by Occam's Razor, that is enough.

If we use Carter's (1983) anthropic arguments and there are 10^9 habitable planets in the Galaxy, then we expect after 10^{10} years that there can be at most 10^8 intelligent civilizations formed, because the probability for forming an intelligent civilization on a habitable planet must be at least an order of magnitude and perhaps many orders of magnitude less than 1, so that the remarkable order of magnitude coincidence between the time for intelligent life to form on Earth and the future main sequence lifetime of the Sun is explained (see the ingenious argument presented by Carter, 1983). Thus approximately 4.5 billion years after metal-enriched stars like the Sun are formed, as these civilizations are forming randomly in an interval of 10^{10} years, we expect the number of civilizations to have formed in our Galaxy by now to be at most 4.5×10^7 and perhaps many orders of magnitude below that. I argue (Gott, 1993) that the probability of a civilization colonizing the Galaxy is less than 10^{-9} (otherwise you would be likely to be a descendant of galactic colonists). Thus, the probability that the Earth would have been previously colonized by extraterrestrials is at most 4.5% and perhaps many orders of magnitude below this. Thus, with Fact B alone and anthropic and Copernican Principle arguments, we can show that it is unlikely that the Earth would have been colonized by extraterrestrials (see also Leslie, 1989, 1990, 1992, who, using similar arguments (discussed by B. Carter in a 1983 talk although never in print, as well as by Nielsen, 1989), also argues that we are unlikely to colonize the Galaxy). Of course, my arguments show that the zoo hypothesis (Ball, 1973) is unlikely as well. If there is a big galactic club, it is improbable for you to belong to one of the few intelligent species that do not belong to it.

My paper (Gott, 1993) indicates that not colonizing is a big mistake we are likely to make. The goal of the human spaceflight program should be to increase our survival prospects by vastly expanding our habitat. But my paper shows that we are not likely to be wise enough to do this. In any case, colonization is not likely to be important, so we can return to using the Drake equation (cf. Drake & Sobel, 1992). My paper contains a 95% confidence level upper limit on the mean longevity of radio-transmitting civilizations $< L > < 12\,100$ years, which has previously been one of the more uncertain terms in the Drake equation. The rate R at which civilizations are forming in the Galaxy is less than $0.01 \ \mathrm{yr}^{-1}$ (i.e. less than 10^8 civilizations per 10^{10} yr) so, by the Drake equation, the 95% confidence level upper limit on the number of radio-transmitting civilizations in the Galaxy now (which equals 95% confidence level upper limit on $R < L >$) is < 121 (Gott, 1993).

I am a strong supporter of SETI. Although I predict that a targeted radio search of 1000 nearby stars is not likely to be successful, I point out that an all-sky radio search could be successful. It is important for any scientific theory to be falsifiable, so doing the targeted radio search of nearby stars provides an important test of my hypothesis that one's position among intelligent observers is not special. Of course, quite aside from that, the search is well worth doing even if its chance of success is low, because it does not cost much to do and if it does succeed, its benefits will be enormous. By showing that colonization is not likely, my paper considerably improves SETI chances over what many people might have thought previously because it allows other intelligent civilizations in our Galaxy. My answer to Fermi's question 'Where are they?' is that a significant fraction of them must be at home, just like us (otherwise, as a random intelligent observer, it would be improbable for you to find yourself sitting on your species' home planet of origin). My 95% confidence upper limit on civilizations transmitting in the radio in the Galaxy now is < 121 (Gott, 1993). This is another test (falsifying my theory if we find more than that). We see more radio pulsars than that in the Galaxy. Also, civilizations could be seen in external galaxies, so all-sky surveys might well succeed. Since Kardashev Type II and III civilizations are likely to be relatively rare (Gott, 1993), we ultimately need surveys sensitive enough for us to detect civilizations like ourselves within the Galaxy and then in external galaxies.

If the luminosity function of transmitters is sufficiently broad, then there is even a possibility that the brightest extragalactic transmitter might be more easily detectable than any in our own Galaxy. Thus, extragalactic frequency standards like those proposed here are profitable even if there are a number of other civilizations visible in our Galaxy. Besides, if the extragalactic frequency standards are sufficiently accurate and easy to use, they might well serve for communication within a galaxy as well. If the probability of intelligent life developing spontaneously is small, then extragalactic searches are the only ones with a chance of success. The points mentioned above show that it is useful to consider strategies to search for life using extragalactic frequency standards. For this purpose, frequency standards like 56.8 GHz and 28.4 GHz tied to the cosmic microwave background radiation are useful.

References

BALL, J. A. (1973). *Icarus*, **19**, 347.
BRACEWELL, R. N. (1975). *The Galactic Club: Intelligent Life in Outer Space*. San Francisco: W. H. Freeman. Distributed by Scribner, New York.
CARTER, B. (1983). *Phil. Trans. Royal Soc.*, **A310**, 347.
COCCONI, G. & MORRISON, P. (1959). *Nature*, **184**, 844.
DRAKE, F. D. & SAGAN, C. (1973). *Nature*, **245**, 257.
DRAKE, F. & SOBEL, D. (1992). *Is Anyone Out There?* New York: Delcacorte Press.
DYSON, F. J. (1960). *Science*, **131**, 1667.
DYSON, F. J. (1979). *Reviews Mod. Phys.*, **51**, 447.
GOTT, J. R. (1980). *M.N.R.A.S.*, **193**, 153.
GOTT, J. R. (1982). *Extraterrestrials, Where Are They?* ed. M. H. Hart and B. Zuckerman, p. 122. New York: Pergamon Press.
GOTT, J. R. (1993). *Nature*, **363**, 315.
GOTT, J. R. & TURNER, E. L. (1976). *Ap.J.*, **209**, 1.
GUTH, A. H. (1981). *Phys. Rev. D.*, **23**, 347.
HART, M. (1975). *Quart. J. Royal Astron. Soc.*, **16**, 128–35.
HART, M. (1982), *Extraterrestrials, Where Are They?*, ed. M. H. Hart and B. Zuckerman, p. 154. New York: Pergamon Press.
HOROWITZ, P. & SAGAN, C. (1993). *Ap.J.*, **415**, 218.
KARDASHEV, N. S. (1964). *Soviet Astronomy – A.J.*, **8**, 217.
LESLIE, J. (1989). *Bull. Canad. Nucl. Soc.*, **10** (3), 10.
LESLIE, J. (1990). *Phil. Quart.*, **40**, 65.
LESLIE, J. (1992). *MIND*, **101**.403, 521.
LINDE, A. D. (1990). *Inflation and Quantum Cosmology*, p. 139, Boston: Academic Press.
MATHER, J. *et al.* (1993). Talk at AAS Meeting, Phoenix, January, 1993.
NIELSEN, H. B. (1989). *Acta. Phys. Polon.*, **B20**, 427.
OLIVER, B. M. (1977). In *The Search for Extraterrestrial Intelligence*, NASA Sp-419, 63.
PENZIAS, A. A. & WILSON, R. W. (1965). *Ap.J.*, **142**, 419.
SANDAGE, A. & TAMMAN, G. (1975). *Ap.J.*, **197**, 265.
SMOOT, G. F. *et al.* (1992). *Ap.J.* **396**, L1.
WEISS, R. (1980). *Ann. Rev. Astron. & Astrophysics*, **18**, 489.
ZELDOVICH, Y. B. (1972). *M.N.R.A.S.*, **160**, 1P.

20

Galactic Chemical Evolution: Implications for the Existence of Habitable Planets

VIRGINIA TRIMBLE

Introduction

We do not know the total set of conditions necessary for the development of sentient life, but it is a pretty safe bet that chemically based life, at least, requires both a wide range of chemical elements and a good deal of time. Other chapters in this volume attempt the difficult task of estimating how many planets might have had long-lived, stable supplies of water, carbon dioxide, etc., and temperate climates. This chapter addresses the much simpler issue of the numbers, ages and locations in our Galaxy of stars with adequate supplies of heavy elements to make terrestrial planets, in principle, possible.

Carbon, oxygen, phosphorus and the other substances needed by terrestrial living creatures have not been here since the beginning of the universe. Rather, they are the products of a long series of nuclear reactions that occur in the centers of (mostly massive) stars and that have timescales of millions to billions of years (Burbidge *et al.*, 1957; Trimble, 1991). In some of the very oldest stars in our Milky Way Galaxy, only one atom in 100 000 is not hydrogen or helium (Edmunds & Terlevich, 1992; Spite, 1992). Enrichment is a continuous process. Even as you read this, massive stars like Betelgeuse and Antares are synthesizing new heavy elements out of H and He, and supernovae like 1987A are spewing out the products to be raw materials for future generations of stars and planets. In astronomical terminology, all the elements except hydrogen and helium are called 'metals'. The reasons are largely historic. At visible wavelengths, most of the strong absorption lines in stellar spectra come from real metals such as iron, calcium and sodium. Oxygen, neon and carbon are actually commoner; but 'metals' they remain.

The particular clump of material that gave rise to the solar system consisted of about 70% hydrogen (by mass), 28% helium and 2% everything else, including (in order of decreasing abundance) O, C, Ne, N, Mg, Si, Fe, S, Ar, Al, Ca, Na,

184

Ni, Cr, P, Mn, Cl, K, Ti, Co, Zn, Cu, and so on down to exceedingly rare substances such as U, Th, Be, Au and Pt (Trimble, 1991). It is probably not an accident that the biologically most important elements (H, CNO, and, to a lesser extent, Na, Cl, P, Fe, Ca, etc.) are among the most abundant. The Earth, of course, has much more than its fair share of metals and much less than its fair share of hydrogen and helium. The resulting solid surface, oceans and air are directly responsible for its home-like qualities.

It is not a coincidence that the solid-surfaced, terrestrial planets are close to the Sun and warm enough for liquid water, while the jovian (gas-giant) planets are in the outer, frigid reaches of the solar system. Common sense and computer models agree that the terrestrial planets of rocks and metals formed where they did precisely because only refractory substances such as iron-magnesium silicates could solidify this close to the new-born Sun. Such a selection effect must occur during the formation of any planetary system. Thus, whatever planets exist in the habitable zone around stars of 0.8–1.2 solar masses will be terrestrial planets. But if the supply of metals is too small, then either no planets at all will form close to the parent star or they will be like Mercury and Mars, too small to hold on to hydrosphere and atmosphere.

How small is too small? No one really knows, because the calculations of planet formation in protoplanetary disks have not been done for compositions different from that of the solar system. But a planet with less than one-third the mass of Earth will have trouble holding on to water and air at room temperature, so it may be reasonable to assume that the probability of habitable planets is significantly reduced for stars with less than one-third the solar fraction of heavy elements.

At the original 'Where are They?' meeting, the scheduled speaker on galactic chemical evolution was the late Beatrice M. Tinsley, for whom I was a last-minute replacement. The influence of her ideas and models remains a living presence in the astronomical community (Tinsley, 1980; Tinsley & Larson, 1977). My interest in extraterrestrial intelligence first arose (as probably for most astronomers of my generation) from a student reading of Shklovskii & Sagan's (1966) *Intelligent Life in the Universe*. It is still a good investment! The listed references include both important introductions like this to the topics of galactic chemical evolution and its implications and specific items cited in the text.

Stellar Populations – the Hopeless Halo

Stars in the Milky Way (and other spiral galaxies) can be divided into about four populations in terms of location, age and chemical composition. The outer halo is nearly spherical, large, but sparsely populated. Its 500 million or so stars are all at least twice as old as the Sun, and very few have more than about 10% of solar metallicity. Intelligent life in the halo could have had 10 billion years more than we have had to think about things, but terrestrial planets may well never have formed.

The inner halo and thick disk populations (Gilmore et al., 1989; Rana, 1991) are somewhat more crowded, somewhat younger and somewhat more metal-rich than the outer halo. Again there would be time (perhaps 10 billion years) for all sorts of things to happen, but very possibly no terrestrial planets for them to happen on.

The halo and thick disk together contain only about 10% of the 10^{11} stars that make up our Galaxy. Most of the rest belong to the thin disk, where the Sun is, and to the galactic central bulge, a region not easy to study, because interstellar dust blocks our view at visible wavelengths. Thin disk stars are, on average, still younger than thick disk stars. Most bulge stars are probably older than the Sun. It is among these populations that we must look for habitable terrestrial planets.

Somewhat surprisingly, the mix of heavy elements found in these various populations is not very different from the solar mix, even when the total amount is much smaller or somewhat larger. The most conspicuous difference is a larger ratio of oxygen (and probably neon, carbon, magnesium and silicon) to iron and adjacent elements when the total metal abundance is low (Rana, 1991; Spite, 1992). The range is about a factor of 4. I cannot see that this matters very much to formation and habitability of terrestrial planets. Trapping metals in silicates and such does not use all the oxygen available in any case, leaving plenty over to make water, CO_2, and so forth. Stars with more serious anomalies, for instance lots of carbon but no oxygen, or the converse, are rare or nonexistent (Trimble, 1991).

The Solar Neighborhood – Are We Normal?

Well-studied stars are mostly ones within 1000 pc or so of the Sun, well away from the galactic center (fig. 20.1). Halo and thick disk stars pass through this local volume, but our primary neighbors, in terms of both numbers and permanence, are other thin disk objects. Most of these are single and binary stars (the latter probably planetless for dynamical rather than chemical reasons). But there are also gravitationally bound clusters of stars and clouds of gas from which new stars are currently forming. The clusters are important because their ages can be measured more reliably than those of isolated stars.

The average metallicity of all thin disk, solar neighborhood objects is about two-thirds of solar. Disconcertingly, the gas clouds now forming stars often seem

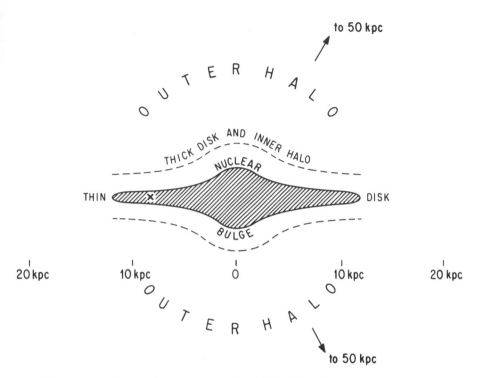

Figure 20.1 Impressionistic view of the Milky Way Galaxy, edge on, showing locations of the stellar populations of the thin and thick disks, nuclear bulge, and inner and outer halos. The Sun is in the thin disk about 8.5 kpc out from the centre at the point marked X. Stars passing through the solar neighbourhood probe the populations of the disks and halos at $R = 8 - 9$ kpc, but not the inner part of the disks or the bulge. The outer halo, as traced by globular star clusters, actually extends at least to 50 kpc, more than twice the size of the drawing. A face-on image of the Galaxy would show spiral arms in the thin disk. A kiloparsec $= 3 \times 10^{21}$ cm.

to be somewhat deficient in heavy elements. This includes the conspicuous Orion Nebula (Peimbert, 1986). If this were the whole story, we might well conclude that the Sun formed from an unusually metal-rich cloud and so has few peers in the planet-forming business. This is not cheating on Copernicus, because we must inhabit a habitable solar system to raise the issue.

The truth is, however, less simple, even locally. First, gas clouds contain a good deal of dust. The dust obviously hides some of the heavy elements (perhaps half) from direct observation in the gas. It also introduces considerable uncertainty into the measurement of metal abundances from atomic emission lines (Henry,

1993). The case for solar anomaly in this sense is, at worst, not proven. Second, while there is a general trend of increasing metallicity with decreasing stellar age, the scatter at each age is rather more than the change of the average over the past 8 billion years or so. Among nearby star clusters, there exist both 8-billion-year-old ones and 1-billion-year-old ones over the full range from one-quarter of solar metallicity to twice it. Similar scatter above and below solar heavy element abundance reveals itself in individual stars of all ages and in gas clouds (Beckman and Pagel, 1989; Rana, 1991). Curiously, Tau Ceti, one of the two objects of the very first SETI project, has only about one-quarter of the solar metal content (G. Smith in Edmunds & Terlevich, 1992), and so may not be a good candidate for a habitable system. Our nearest neighbor, Alpha Centauri, on the other hand, has about twice solar metallicity (Neuforge, 1993), and somewhat larger age as well.

In summary, then, the solar neighborhood contains some stars whose age and heavy element abundance are both at least as large as those of the Sun. They are not, however, the majority. A picture of galactic chemical evolution based only on local data might look like fig. 20.2 and lead to the conclusion that, indeed, terrestrial planets more than 5 billion years old are probably quite rare.

The Bulge and Inner Disk

The situation looks quite different when we consider the half or so of galactic stars nearer the center than we are. In the Milky Way (and most other massive galaxies, both spiral and elliptical), the several populations all display gradients of metallicity, rising toward the center (B. J. E. Pagel in Mathews, 1988). These gradients probably have several causes operating at once, including earlier and more vigorous star formation in the denser central regions, more complete retention of enriched material expelled by supernovae there, and (perhaps) inflow of material enriched by early generations of stars elsewhere in the Galaxy.

The result is that the average K giant in the galactic bulge has twice the solar iron abundance (Frogel, 1988), with considerable scatter in both directions. Main sequence stars like the Sun are too faint to be studied directly in the bulge region, but there is no reason to think that they will be chemically different from the giants. Many bulge stars are also very old. It is as if the Galaxy (or at least its flattened parts) had formed from the inside out (J. M. Scalo, private communication, 1992), with disk star formation beginning at the dense center about 12 billion years ago, and gradually working its way out to reach the solar neighborhood 8–10 billion years ago (Winget et al., 1987). The total chemical evolution of the Milky Way therefore looks more like fig. 20.3 than like fig. 20.2.

In other words, even after removing all the halo and thick disk stars, the younger

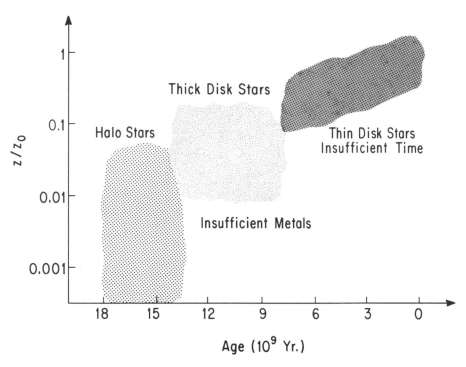

Figure 20.2 The age-metallicity relation for stars in the solar neighbourhood. The vertical axis gives metal abundance by mass, z, in units of that of the Sun $z_0 = 0.017$. Most halo and thick disk stars are probably too deficient in heavy elements to have formed terrestrial planets, while most of the metal-rich, thin disk stars are younger than the Sun, allowing insufficient time for life to develop.

In other words, even after removing all the halo and thick disk stars, the younger and more metal-poor stars of the outer thin disk and all the binaries from the inventory, the Milky Way probably still contains at least 10^{10} stars that could have harbored habitable, terrestrial planets for more than 5 billion years. There is, of course, no guarantee that any of them actually do so. In addition, the majority of hospitable sites are likely to be nearer the galactic center than we are, a consideration that might or might not be of any use in formulating search strategies. The distribution of old terrestrial planets will certainly matter for travel or communication plans. A round trip to the galactic center takes more than 55 000 years; and half the good planets will be on the other side.

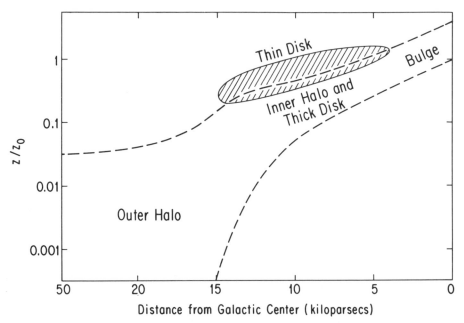

Figure 20.3   Gradients of metallicity vs. radius in the spheroidal and disk populations of the Milky Way. Stars considerably older than the Sun occur in all three spheroidal components, outer and inner halo and nuclear bulge.

Conclusions

Chemically based life requires both generous supplies of carbon, oxygen and other heavy elements and long periods of time for its development. Because the necessary elements form gradually in stellar interiors, and the universe is only 2–4 times as old as the Earth, one might suppose that life is rare simply because the combination of sufficient time and sufficient heavy elements is rare. This seems not to be the case. It is, however, probable that most of the stars that are both rich enough in metals to have terrestrial planets and more than about five billion years old are considerably closer to the center of the Galaxy than we are.

References

BECKMAN, J. E. & PAGEL, B. J. E. (eds.) (1989). *Evolutionary Phenomena in Galaxies.* Cambridge: Cambridge University Press.

BURBIDGE, E. M., BURBIDGE, G. R., FOWLER W. A. & HOYLE, F. (1957). *Reviews of Modern Physics*, **29**, 547.

EDMUNDS, M. G. & TERLEVICH, R. J. (eds.) (1992). *Elements and the Cosmos*. Cambridge: Cambridge University Press.

FROGEL, J. A. (1988). The galactic nuclear bulge and the stellar content of spheroidal systems. *Annual Reviews of Astronomy and Astrophysics*, **26**, 51.

GILMORE, G., WYSE, R. F. G. & KUIJKEN, K. (1989). Kinematics, structure, and chemistry of the Galaxy. *Annual Reviews of Astronomy and Astrophysics*, **27**, 555.

HENRY, R. C. B. (1993). *Monthly Notices of Royal Astronomical Society*, **261**, 311.

MATHEWS, G. J. (ed.) (1988). *Origin and Distribution of the Elements*. Singapore: World Scientific.

NEUFORGE, C. (1993). *Astronomy and Astrophysics*, **268**, 650.

PEIMBERT, M. (1986). In *IAU Symp. 115, Star Forming Regions*, ed. M. Peimbert & J. Jugaku. Dordrecht: Reidel.

RANA, N. C. (1991). Chemical evolution of the Galaxy. *Annual Reviews of Astronomy and Astrophysics*, **29**, 129.

SHKLOVSKII, I. S. & SAGAN, C. (1966). *Intelligent Life in the Universe*. San Francisco: Holden-Day.

SPITE, M. (1992). In *IAU Symp. 149, The Stellar Populations of Galaxies*, ed. B. Barbuy & A. Renzini, pp. 123–32. Dordrecht: Reidel.

TINSLEY, B. M. (1980). *Fundamentals of Cosmic Physics*, **5**, 287.

TINSLEY, B. M. & LARSON, R. B. (eds.) (1977). *The Evolution of Galaxies and Stellar Populations*. New Haven: Yale University Observatory.

TRIMBLE, V. (1991). *Astronomy and Astrophysics Reviews*, **3**, 1.

WINGET, D. E. *et al.* (1987). *Astrophysical Journal*, **315**, L77.

21

The Frequency of Planetary Systems in the Galaxy

JONATHAN I. LUNINE

Planetary Systems and Life

One of the most profound questions which we as a species can ask is whether we are alone in the cosmos. Posed in a slightly more specific fashion, how many worlds, N, in our Galaxy are inhabited by technological civilizations? One way to estimate this number is to express it as the product of a series of probabilities, e.g. (Shklovskii & Sagan, 1966)

$$N = R_{star} f_p n_e f_L f_i f_c t_{span} \qquad (21.1)$$

Here R_{star} is the rate of star formation in the Galaxy, averaged over the lifetime of the Galaxy; f_p is the fraction of stars with planetary systems; n_e is the mean number of planets in each planetary system with environments conducive to the formation of life; f_L is the fraction of such planets on which the formation of life is actually consummated; f_i is the fraction of such biospheres which evolve intelligent life; f_c is the fraction of such intelligent species which develop technological civilizations (capable of communicating); and t_{span} is the average lifetime of such technological civilizations.

The star formation rate R_{star} is a function of stellar mass and may vary spatially from one star-forming region to another, as well as temporally; nonetheless, the problem of its determination is amenable to current observational techniques (Zinnecker *et al.*, 1993), and we seek to know it only to an order of magnitude or so (the integrated function we are really after, the number of mid- to late-type main sequence stars in the Galaxy, is well known and of order 10^{11}). The parameters n_e, f_L, f_i, f_c and t_{span} are at present completely intractable from an observational point of view, and except for the first cannot even be approached on the basis of physically sound theory (arguably so for f_L). Only f_p, the fraction of stars with planetary systems, stands at the interesting threshold of being nearly accessible with current observational techniques such that one can anticipate significant discoveries (positive or negative) in the coming decade. It has been possible for

192

the past decade to observe indicators of planetary system formation around proto-stellar objects (Black & Matthews, 1985) with very rapid advances in capabilities and discoveries over the second half of that period of time (Levy & Lunine, 1993). Theoretical work on how planetary systems form and with what frequency extends back, in essence, to Laplace (1796); however, the advent of very high speed computing capabilities during the decade of the 1980s has led to major strides in theoretical understanding which are as dramatic as the progress of observations.

The present chapter, then, summarizes current understanding of how planetary systems form and the observational evidence that such systems might be common. It then goes on to describe existing techniques for searching for mature planetary systems. Because of the book's theme, we limit ourselves to objects with a few Jupiter masses or smaller. Objects in the range 10–80 Jupiter masses, often called brown dwarfs, may form singly or in binary systems in the same fashion as do stars, even though they do not fuse hydrogen. The upper mass limit for forming giant planets as part of a planetary system is uncertain (Black, 1986), but current models suggest 10 Jupiter masses to be a reasonable upper limit.

Gauging the Frequency of Planetary Systems from Precursors

In the absence of direct information on the incidence of planetary systems, one can do little other than use observations and modeling of the formation of planets to either validate or contradict the notion that planets are a natural consequence of the star formation process. Structures likely to be associated with planet forma-tion, such as disks of gas and dust around newly forming ('proto-') stars, are much easier to observe than planets themselves because of the enormously larger surface area over which light is scattered, absorbed or reflected by such disks. An optically thick disk 30 astronomical units (AU) in radius and 1 AU in thickness has a radiating surface area 3×10^7 times larger than the total surface area of the planets in our solar system. (One astronomical unit, or AU, is the Earth–Sun distance.) Solar composition disks, composed largely of hydrogen and helium gas, are self-luminous because they dissipate the energy of infalling material through turbulent motions over a range of size scales, leading to temperatures extending from several thousand K near the inner boundary to 50 K or less a few tens of AU from the disk's center (Wood & Morfill, 1988). The bulk of the disk surface radiates in the near- to far-infrared, but with a significant component also in the optical. As quantified in the next section, planetary bodies up to ten Jupiter masses or so quickly lose their internal heat of collapse, and can be seen over most of their lifetime by the reflected light of the parent star, or by observations of their own luminosity in the mid- to far-infrared. Disk precursors to planetary systems are much easier to observe than mature planets themselves.

The arguments for disks being the signal precursor of planetary systems lies in the geometry of our own solar system, and in the apparently high frequency with which disk-like structures are observed in association with stars which are so young ('pre-main-sequence', or PMS stars) that they have not moved onto the so-called main sequence of stars which undergo hydrogen fusion. Our solar system is characterized by a single central star bearing most of the mass of the system (99.9%) but less than 3% of the angular momentum (Smith & Jacobs, 1973). The remainder of the angular momentum lies in the orbital motion of nine planets, their moons and an assortment of small bodies (such as asteroids and comets which together comprise as much as 100 Earth masses of material: Mumma *et al.*, 1993). Most of this material moves in the same direction – counterclockwise when looking down on the Sun – and, more remarkable, seven of the nine planets have nearly circular orbits which are inclined no more than several degrees relative to one another. The most dynamically plausible origin for this kind of structure is an assemblage of gas and dust, flattened into the form of a disk as the material collapses to a common center and is forced to conserve its initial angular momentum content.

The formation and subsequent evolution of such disks are much better understood now than even five years ago, with the advent of powerful new observational and computational tools. While the beauty is in the details, it is possible to attempt a consensus picture of the overall process. We paraphrase and extend here the excellent summary of Adams & Lin (1993): Star formation occurs within 'core' regions of molecular clouds, which possess higher-than-average density of gas and dust. These regions are believed to be at least partially supported through the molecular cloud magnetic field which acts on the ion population of the cloud and thereby creates a pressure in the bulk gas by ion–neutral collisions. By virtue of their higher density cores tend to have a lower ion concentration. The magnetic field slowly diffuses outward, increasing the central density of the cores, further decreasing the ion population and, hence, magnetic support. This process continues until the core is supported against self-gravity largely by thermal pressure, an unstable quasiequilibrium state which is the starting point for dynamical collapse.

As the core collapses, some material forms a small, hydrostatic object at the center, the 'protostar'. Much of the material forms a disk around the protostar by virtue of the conservation of the angular momentum possessed by the collapsing core. Infall of material continues at a high rate, mostly into the disk, which is comparable in mass to the protostar. Gravitational instabilities and/or strong magnetic fields create torques which tend to move mass inward to the protostar and angular momentum outward toward the edge of the disk. The mass and luminosity of the protostar increase, and eventually it forms a wind which flows preferentially outward along the rotational axis of the system (Adams & Lin, 1993). Efficient inward processing of disk material to the protostar and eventual depletion of infalling material reduce the mass of the disk, and further evolution of mass and angular momentum through the disk is by less efficient mechanisms such as large-scale turbulence (driven by thermal convection or large-scale circulation patterns). The longest stage in the evolution of the disk, the viscous accretion phase, is plausibly the period when gas-rich planets form. The stellar wind gradually widens in angular extent over time, separating the star from the disk and eventually dissipating the disk. Systems at this stage are now recognized as classical T Tauri stars; so-called weak-line T Tauri and naked T Tauri stars may be systems in which the disks are in advanced stages of dissipation (or in which no disk was ever present; see Basri and Bertout, 1993). The central star continues to evolve toward stable hydrogen burning and onto the stellar main sequence.

On the basis of observations and stellar evolution models, T Tauri stars range in age from 10^6 to 10^7 years, with naked T Tauris tending to be older (Stahler & Walter, 1993). Few optically thick disks have been observed around stars with model ages exceeding 1×10^7 years, and none around stars with ages exceeding 3×10^7 years (Strom et al., 1993), which sets a limit then on the lifetime of massive, gas-dust disks (a fraction of the dust component may survive throughout much of the main sequence lifetime, as discussed below). These age limits pose very tight constraints on gas-giant planet formation, as we discuss in the following section.

Observations of active star-forming regions reveal evidence for disks of gas and dust in two ways (Strom et al., 1993). First, broad forbidden line spectroscopic profiles are diagnostic of outflows from PMS stars. A significant fraction of such emissions show only the blueshifted component; redshifted emission from a receding component is absent, and the simplest interpretation is that an optically thick disk is obscuring this far side component over radial scales of ten to hundreds of AU from the central star. Second, a broad assortment of observations from infrared to millimeter wavelengths reveal excess emission, relative to that expected for the long-wavelength tail of the stellar blackbody radiation curve, which is most readily interpreted in terms of disks which are both reradiating stellar energy and emitting photons associated with heating of the disk by accretional processes. Recent interferometric maps at millimeter wavelengths have succeeded in directly resolving, usually in emission of heavy isotopes of carbon monoxide, disk-like features of extent ~ 1000 AU around a few PMS stars (Beckwith & Sargent, 1993). These identifications bolster the interpretation of disks as being responsible for the excess emissions seen around the much larger number of PMS stars which can be surveyed using noninterferometric techniques.

Roughly 30–50% of PMS stars of mass less than 3 solar masses show such evidence for optically thick disks, as do an as yet poorly known fraction of more massive stars (Strom *et al.*, 1993). Inferring masses for these disks requires assuming knowledge of the dust-to-gas ratio in the disks, since the dust is largely responsible for the emission, while the gas dominates the mass (however, the millimeter-wave observations of molecular emission do sample gas emissions). Using the dust-to-gas ratio one would get from elemental abundances in our Sun, disk masses typically exceed 0.01 solar masses (Strom *et al.*, 1993). Taking all of the material currently in the planets of our own solar system and augmenting it with enough hydrogen and helium to yield the elemental abundance in the Sun gives a mass for the disk out of which our own system formed of roughly 0.01 solar masses. Thus, disks inferred or observed around very young stars in active star-forming regions apparently have the mass necessary to produce planetary systems. Moreover, the size scale of such disks are within one or two orders of magnitude of the size scale of our own solar system (excluding the outer cometary Oort cloud, which has extent 100 000 AU, and which is most likely composed of objects gravitationally ejected from the original, more compact nascent solar system of extent perhaps 100 AU; see the discussion in Weissman, 1990).

The formation of planets can therefore plausibly be linked, albeit indirectly, to the presence of gas and dust disks around forming stars. One might therefore predict the incidence of planetary systems by determining the fraction of stars which possess disks at some time during their earliest evolution. By this measure, perhaps one-half of stars with masses three times or less that of the Sun should have planetary systems. Before jumping to this conclusion, however, one must ask whether planet formation is necessarily the end result of all gas disks around pre-main sequence stars. In fact, the answer is no; binary star formation is another possible consequence of disk formation.

Perhaps 50–80% of main sequence stars are binary, including those multiple systems in which a more distant third star is present as a captured companion; hence, binarity is the usual outcome of star formation (see the review by Bodenheimer *et al.*, 1993). Although capture of a single companion into a long-lived close orbit is possible, it requires very high densities of stellar neighbors or special conditions; hence, the preponderance of binaries tells us that the mechanisms of star formation associated with collapse of the source molecular cloud core often, or usually, produce binary stars. At issue for us is whether binary star formation occurs through, or instead of, disk formation. Plausible models for binary star formation include direct fragmentation during collapse of the protostellar material, and formation of a disk followed by fragmentation of that structure.

Direct fragmentation of the collapsing protostellar material readily yields binary stars with separation of order 100 AU, but has not been convincingly demonstrated as a mechanism for forming close binaries (Bodenheimer et al., 1993). Therefore fragmentation of a gaseous disk is implicated as a mechanism for forming at least the close binary systems. Formation and subsequent fragmentation of disks to form binaries requires a critical mass of the disk relative to that of the central object. For a disk mass fraction (relative to the total) of order 0.1 or larger, the disk is unstable to gravitational instabilities, but this critical number depends upon the disk angular momentum as well (Bodenheimer et al., 1993). For stars like the Sun, the critical mass is of order 10 times the minimum mass needed to produce a planetary system similar to our own; it would therefore seem that there is plenty of mass range within which the disk has sufficient material to make a robust planetary system but not enough to form a binary. However, it is unclear whether the collapse of cloud cores to produce solar mass stars can yield relatively low-mass disks without first passing through a stage in which the disk mass is comparable to the protostar. During this stage gravitational instabilities may move mass rapidly into the protostar and angular momentum out to the edge of the disk, leading to a lower-mass disk as the bulk of the accretion subsides (Adams & Lin, 1993). It is not clear yet what fraction of disks survive this early stage to become planet-forming, as opposed to fragmenting into a binary system.

It is therefore impossible to say, at present, what fraction of disks survive to a stable phase during which planet formation can occur. However, the observational test suggested by the theory is to determine what fraction of pre-main sequence stars have low-mass disks (i.e. less than 0.1 of the protostellar mass). Such disks are likely candidates for stability against catastrophic fragmentation and binary star formation.

Main sequence stars also possess disks. A disk has been directly imaged around β Pictoris in the optical (Smith & Terrile, 1984), which also shows characteristic infrared excesses which were detected by IRAS during its infrared survey mission in 1983. By virtue of the β Pictoris identification, over 100 other main sequence stars with similar excesses are strongly suspected of having such disks as well (see the review by Backman & Paresce, 1993). Analysis of the emission reveals it to be essentially all continuum thermal radiation, with little or no contribution from line emission. Therefore these disks are dominated by dust grains, with little or no gas. Typical grain sizes for three of the best-observed disks range from 1 to 100 μm, suggesting that these grains are partially evolved upward from interstellar grain sizes (Backman & Paresce, 1993).

Do these disks of small particles represent indicators of planetary system formation, or are they debris from stars which failed to form planets? The mass of dust inferred from the disk is very sensitive to the assumed size distribution, since most of the mass can be in large grains which are sufficiently rare that they do not contribute to the emission; estimates range from $\sim 10^{-3}$ to 10^4 Earth masses (Backman & Paresce, 1993). A tantalizing but indirect indication of planet formation is the maximum grain temperature inferred from emission at different wavelengths observed by IRAS, which sets an inner disk boundary between 20 and 70 AU for the three best-measured disks (though it is impossible to tell how sharp these boundaries are). One interpretation is that these boundaries reflect dynamic clearing of grains by a planetary system whose extent corresponds to the outer boundary of the cleared zone (Diner & Appleby, 1986). Another explanation for the cleared regions is sublimation of the grains inward of the boundary. If the cleared region is due to planets, then the analog to these dust disks in our own solar system is the hypothetical Kuiper belt of material which, on the basis of a variety of evidence, contains material of cometary size and smaller in a disk-shaped zone between 30 and ~ 100 AU from the Sun (Duncan et al., 1988). Loss processes for small particles in the dust disks argue that they are replenished by collisions among larger objects, e.g. comets (Zuckerman, 1993).

The best-observed cases of the dust disks are seen around A-type stars which have been on the main sequence for times ranging from 0.4 to 2 billion years (Backman & Paresce, 1993), all of which ages are longer than the estimated timescales for forming planets. The other ~ 100 stars with infrared excesses interpreted to be disks range in spectral type from B down to K (the last being cooler and less massive than our own Sun). Ground-based surveys hint that such excesses are common, with perhaps half of A-type stars and 10% or so of cooler stars possessing such signatures. The surveys suggest that main sequence stars fairly commonly end up with condensed, rocky and/or icy material in orbit about them. However, not all (and maybe not most) dust disks necessarily have a geometry suggesting embedded planetary systems. Debris clouds are found in the IRAS survey around stars which are members of multiple systems, and at least one well-studied case involves a massive dust disk orbiting a few AU from a star which has stellar companions at 1 and > 20 AU (Zuckerman & Becklin, 1993). Therefore the abundant presence of dust disks does not by itself serve as an indicator that stable planetary systems are common features of stars; however, particular geometries such as a cleared inner zone seen in a disk around a single star may be indirect evidence that planets are present.

There is also indirect evidence for planets around post-main sequence stars with outflows; roughly 80% or more of these outflows show evidence for non-spherical symmetry. The most popular model to explain the outflow geometry requires a close binary system; such systems, however, are not sufficiently common to account for the very high percentage of flows which are asymmetric (Zuckerman, 1993). An alternative is that Jupiter-like planets are present close to the star in some cases, and that these ablate, forming a disk which partially collimates the flow from the central star, or even triggers SiO masers in which the source material is ablated from silicate satellites of the planet (Struck-Marcell, 1988). If these interpretations are correct, then by inference a significant fraction of stars, i.e. of order 10%, must have Jupiter-like planets.

In summary, while gas-dust disks appear to be common features around proto-stars, the number of such disks which are capable of forming planets is uncertain. The high incidence of binarity around main sequence stars, and the evidence for fragmentation of relatively massive disks in hydrodynamic simulations, allows for the possibility that the usual outcome of disk formation is creation of binary stars. However, it is just as likely, given our current state of ignorance, that most disks evolve rapidly through an early massive phase, shedding mass onto the central protostar and ending up as dynamically stable, slowly evolving low-mass disks which retain enough material to form planets. The discovery of dust disks around main-sequence stars is not by itself support for the latter picture, except that the mass of dust, age of the disks and possible presence of cleared regions close to the star all argue for evolution toward larger bodies, perhaps as large as planets. Gaseous disks which are 1–10% of the mass of their central PMS star would seem to be very good candidates for planet formers; systematic surveys which could identify such disks are feasible but remain somewhat model-dependent because the predominant constituent, molecular hydrogen gas, is not measured directly. Thus, while surveys of proto-stellar systems may be useful in indirectly determining the frequency of planetary systems, they are not a substitute for direct detection of planetary systems themselves.

Taxonomy of Planetary Systems

Before proceeding to describe search techniques, we must consider what we are looking for. Search techniques are often evaluated on the basis of assuming planet-ary systems akin to our own solar system. How common, though, is our own planetary system? This question can only be answered at present by assessing the likelihood that the circumstances and mechanisms for forming the planets in our own solar system might be commonly repeated around other stars. The flaw in this approach is that difficulties in understanding particular types of objects could lead, perhaps too quickly, to the notion that those objects are rare in other planetary systems. As we shall see in this section, such is the case currently for the giant planets, an unfortunate state of affairs, since current technologies force search strategies to hinge on the frequency of giant planets being some measure of the frequency of planetary systems in total.

Early expectations that planetary systems might commonly contain a mix of giant and terrestrial planets are based on very simplified simulations of planet formation (Dole, 1970; updated somewhat by Isaacman & Sagan, 1977). Such models contained a number of approximations in terms of initial conditions, assumptions about size distribution of 'feeder nuclei', distribution of orbital eccentricities of material, the physics of gas accretion and timescales. In fact, the problem of accumulation divides itself into two problems: the growth of grains up to a kilometer in size, below which coupling to the gas of the pre-planetary disk strongly affects the motions, and the agglomeration of larger bodies up to planet-sized, during which gravitational interactions and collisions can potentially pump up orbital eccentricities leading to a complex situation of accretion, disruption and ejection of material from the system. The initial problem, production of kilometer-scale 'planetesimals' from planets, is now better understood in terms of microphysical coagulation processes in a turbulent, gaseous disk (Weidenschilling & Cuzzi, 1993). Although extremely rapid formation of kilometer-sized condensates by gravitational instability at the disk midplane (Goldreich & Ward, 1973) is no longer considered likely, owing to turbulent spreading of the dust vertically, other microphysical processes can lead to growth of meter-sized objects on timescales much less than the lifetime of the disk estimated above. Growth to kilometer-size is somewhat more problematic, but is thought to be possible on short timescales as well (Weidenschilling & Cuzzi, 1993).

Growth of planets from planetesimals further divides itself into two problems, and it is here that the uncertainties arise in the ease with which giant planets can form. The terrestrial planets of our own solar system formed in a region close enough to the Sun that the mean velocity between growing planetesimals, of order a few kilometers per second and determined in large part by the escape velocity from the largest objects, was much less than the escape velocity from the solar system, 13 km s^{-1} at 1 AU (Wetherill, 1993). Therefore, this system was gravitationally tightly bound, favoring efficient accretion of material into progressively fewer objects of increasing size. Moreover, timescale constraints on this process are weak: since the atmospheres of the terrestrial planets are likely to be secondary products of outgassing of trace volatiles, or addition of material from impacting comets early in the history of the solar system, the lifetime of the gaseous disk (10^7 years) is not relevant. Other detailed arguments involving the precise geometry of the terrestrial planets and asteroids, the role of Jupiter in perturbing planetesimal orbits and chemical evidence provide some constraints, but for our purposes these are not terribly interesting. Given a solar mass star, and a 0.01 solar mass gas-dust disk out of which silicates eventually condense and/or agglomerate, current simulations of accretion indicate that some number of rocky bodies will readily form on timescales of 10^8 years or less. This is likely to be the case over a range of stellar and disk masses, for approximately solar composition material.

The giant planets present a different problem. The observed enhancements of 'z-elements' (those heavier than hydrogen and helium) in these objects relative to solar abundances are very well established from modeling of their gravitational fields as well as direct atmospheric measurements (e.g. Zharkov & Gudkova, 1991; Podolak et al., 1993). Models in which giant planets form by gravitational instability directly from the gas cannot explain the chemical compositions of the giant planets, and do not work dynamically unless the disk is very massive, in which case the net result is binary star formation (see above). Theorists today understand giant planet formation in terms of initial accumulation of rock and ice planetesimals to form a terrestrial-planet-sized core, which progressively attracts gas gravitationally from the surrounding disk. Eventually the mass of the object becomes sufficient to enable accretion of gas to increase rapidly, leading essentially to a hydrodynamic collapse of gas onto the core, even as additional rock and ice planetesimals fall into the assemblage, vaporizing and contributing z-element materials to the gaseous envelope. Multiple numbers of these gaseous protoplanets may then accrete together to form a giant planet (Stevenson, 1984), or the rapid accretion of gas and solids may continue on a single, largest embryo until the giant planet is formed (e.g. Podolak et al., 1993).

The constraints and potential obstacles to giant planet formation in this physically appealing picture are several. Most important is the timescale issue: the gaseous disk must be present to supply the envelopes of the giant planets, and this sets a formation time of order 10^7 years or less. However, orbital timescales in the outer disk are longer than in the region of the terrestrial planets and, hence, accretion times are longer. An additional inefficiency in the process arises from the fact that escape velocities from the outer solar system are low, 4 km s^{-1} at 10 AU, comparable to the mean planetesimal speed determined by the escape velocities of the largest, growing cores (Wetherill, 1993). Therefore, ejection of material from the solar system on hyperbolic orbits becomes an increasingly likely outcome of planetesimal encounters as one moves outward in the solar system, rendering accretion less efficient and timescales longer.

Early models led to disastrously long timescales for core accumulation, exceeding 10^9 years for Uranus and Neptune and potentially too long even at the jovian region of formation. However, proper accounting of the energy partitioning among planetesimals leads to very rapid growth of the largest planetesimal at the expense of the smaller. This process, runaway accretion, leads to the formation of giant planet cores on sufficiently short timescales, provided that the surface density of solids in the outer disk is enhanced by a factor of $3+$ over that predicted by the minimum mass disk considered above (Lissauer & Stewart, 1993). Two ways to do this are to increase the mass of the entire disk, gas and dust, or to selectively enhance the grain component. Stevenson & Lunine (1988) argued that formation of Jupiter was favored by being just beyond the threshold at which water ice could condense out in the disk, since water vapor moving diffusively through the gas would be cold-trapped out at the condensation boundary, enhancing the surface density of solids. However, this process requires some restrictive conditions to yield enough ice, and works only for Jupiter.

Certain astrophysical observations and evolution models argue for the existence of disks more massive than the minimum mass disk as computed in the previous section, but this argument somewhat narrows the range of acceptable disk masses in which giant planets can form: simulations show disks to be unstable above a mass of 10% that of the parent star, and, hence, must either evolve rapidly to less massive disks or break up to form a binary star (see previous section). The range of disk masses set by the lower limit of rapid giant planet formation and by the upper limit of stability could be less than a factor of 10. It is highly conceivable that many disks evolve rapidly through their massive stage to a mass which is too small to achieve giant planet formation before the nebular gas is fully dissipated. Formation of giant planets closer to the parent star than in our own solar system shortens encounter timescales, but the amount of solid material drops by a factor of 10 inward of the radial distance at which water ice condensation can occur. Since the surface density of grains in protoplanetary disks is a very weak function of radial distance (except where condensation occurs), the loss of stable water ice as a core-forming material mitigates against the advantages of moving the process inward.

Other unanswered questions about giant planet formation exist as well. In particular, the dynamics of the gas accretion around the core is as yet not well understood; very-high-fidelity hydrodynamic models suggest the onset of an instability during core collapse which could force some of the material to expand away from the forming planet (Wuchterl, 1990); perhaps accretion of the gas is a multistage, inefficient process which does not always yield massive gas giants. The much smaller amounts of gas possessed by Uranus and Neptune may reflect timescale issues, or the delicate nature of the process of collapse of gas around cores.

In summary, while accumulation of solids to form terrestrial-sized planets appears to be a robust process, the same cannot be said for giant planets at this point in our understanding. Formation of gas giants may require a more restrictive range of disk mass than gas-free planets, and therefore negative results of searches which are sensitive only to Jupiter-mass objects may underestimate the number of planetary systems. The absence of giant planets from a planetary system might not bode well for intelligent life. Wetherill (1993) points out that the giant planets of our own solar system cleared out a significant amount of material by gravitational interactions. In their absence, a much larger Kuiper belt of material could have supplied to the Earth over geologic time a flux of comet-sized impactors 1000 times larger than our planet is estimated to have actually suffered. Catastrophic, worldwide ecosystem changes on timescales of 10^5 years or less, instead of the 10^8 years seen in the geologic record, might have eliminated the possibility of the development and evolution of complex life forms. By this argument, only those planetary systems which contain terrestrial planets and gas giants are of interest in evaluating eq. (21.1); pure terrestrial planet systems may have f_i (if not f_L) $<<1$.

Finally, we briefly note a different planet formation process, that associated with blowoff of material during supernova explosions or other end-stage stellar processes. The inferred presence of at least two planetary mass companions (≥ 3 Earth masses each) orbiting a neutron star ('pulsar planets') (Wolszczan & Frail, 1992; Wolszczan, 1994), discussed further in the subsection on indirect searches, can best be understood in terms of formation as a consequence of the supernova explosion of the neutron star progenitor, rather than in terms of pre-existing companions which wholly or partially survive the explosion (Bodenheimer, 1993). The most plausible formation mechanisms involve a disk, created either from material which falls back from the explosion (Lin et al., 1991) or from remains of a companion object (Podsiadlowski et al., 1991).

In either case we must ask whether the presence of these planetary mass companions has implications for the theme of this book, the incidence of extraterrestrial life. The direct question, whether life could conceivably occur on any planet formed around a neutron star, has yet to be thought through. The lack of hydrogen in the supernova ejecta which eventually form such a planet is a severe problem, since it implies an absence of water and hydrocarbons in spite of the high abundance of oxygen and carbon. One might on that basis argue that $n_e <<$ 1 in eq. (21.1), for such objects. Such speculations are, however, far less interesting than the simple existence of such planets. If indeed they are a product of the end phase of a star's evolution, rather than a remnant of pre-existing planets, they demonstrate nature's ability to form planetary-scale objects under at least two very different astrophysical situations (Levy, 1993). This may bolster the confidence of the theoretician that planet formation is a reasonably robust process, and give the optimist cause to argue *against* a very small value of f_p in eq. (21.1).

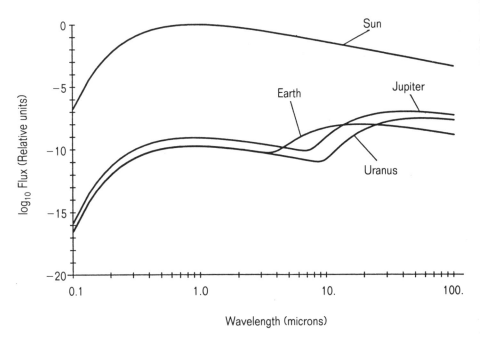

Figure 21.1   Planck curves for three planets and the Sun, expressed as the logarithm of the flux (relative units), from Burke (1992). Both reflected and thermally emitted photons are included for the planets.

Direct Searches

We focus in this and the next section on specific techniques for searching for planets. Much of the discussion is based on material drawn from two published reports, one by the National Academy of Sciences (COMPLEX, 1990) and the other by NASA (Burke, 1992).

Direct detection techniques are conceptually simple; one attempts to measure photons reflected or emitted from the planet against the background of the parent star. Figure 21.1, from the NASA report, illustrates the great difficulty associated with this technique. The figure compares on a logarithmic scale the Planck functions derived from reflected light and thermal flux from Earth, Jupiter and Uranus (at their appropriate semimajor axes) with the Planck function computed from the solar flux. The calculation for the thermal flux involves determining the effective temperature, or photospheric temperature, from the actual flux, and from that computing the Planck function. The reflected light is the solar flux diluted by the distance and relative disk sizes of the Sun and planets. In the optical to near-infrared, planetary fluxes are as much as 9 orders of magnitude less than the solar flux. Earth outshines Jupiter in the 5 μm region, by virtue of its higher effective temperature, and is virtually identical with Uranus in the reflected light region of the spectrum. Far out in the infrared, beyond the Planck function peaks, Jupiter and Uranus are only 6 or so orders of magnitude less bright than the Sun. Planetary emission features are not sufficiently bright to significantly enhance the ratio, though nonthermal radio emissions from the jovian magnetosphere may exceed solar radio noise at times. Whether one could identify such outbursts as planetary in character when observing other solar systems is unclear. (Of course, detection of the radio signature of an intelligent extraterrestrial species would be assumed automatically to imply the presence of a planetary system!)

An additional consideration is the self-luminosity of giant planets. Jupiter has an internal heat source which contributes roughly as much infrared energy as does sunlight thermalized in its atmosphere, but this internal contribution was larger earlier in its history. Many models of giant planet evolution have been published, which for our purposes yield similar results. The current age of Jupiter is 4.5×10^9 years; at an age of 1×10^9 years the luminosity was roughly triple that of today (Saumon et al., 1992). Doubling the mass of Jupiter yields a luminosity ~ 5 times higher at the present age, while keeping the radius (and, hence, the reflected light from the central object) constant (Hubbard, 1977). Thus, direct detections might be easier for more massive planets or earlier times (unfortunately, the easiest way to locate younger stars and, hence, planets is to look at earlier spectral types, for which the central star's luminosity is of course higher). Terrestrial planets of order the mass of the Earth are not self-luminous to an extent of interest to us here and must always be detected by reflected light of the parent star.

Instrument strategies for the daunting task of directly detecting planets are detailed in the reports cited above, and will only be listed here. These include large apertures (for high spatial resolution), high-quality optics and mountings to minimize scattered light, apodizing tapers and coronographs to reshape or reduce the diffraction sidelobes of the central star, interferometers to greatly enhance the spatial resolution and, hence, reduce the sidelobe problem, and adaptive optics to reduce or eliminate wavefront errors associated with turbulence.

The necessity for such techniques is well illustrated by COMPLEX's (1990) analysis of HST's ability to detect Jupiter orbiting a solar type star with 5 AU separation, the system assumed to be 10 pc from the Earth. At an assumed search wavelength of 0.25 μm Jupiter's flux is 10^{-9} that of the solar luminosity parent. For diffraction limited optics (no spherical aberration) HST's image of a point source, the central star, is an Airy diffraction pattern with the first dark ring at a radius of 0.026 arcsec. The image of Jupiter lies at 40 times the Airy radius of the stellar image, and the consequent ratio between planetary and stellar flux is 10^{-5}. Detection of this small a ratio is not feasible for HST; the faint object camera would need to accumulate 10^{10} photoelectrons per resolution element to achieve a 10^5 signal-to-noise ratio, which in turn would require unrealistically long observation times.

Moving the Jupiter-type planet farther from the parent does not work, because at the assumed wavelength the reflected sunlight falls off too rapidly to allow advantage to be taken of the reduction in the stellar sidelobe pattern. An increase in mass, and, hence, the effective temperature, is expressed mostly in the mid-infrared, where the peak of the Planck function is located. Increasing the reflectivity of the atmosphere yields a factor of 2 increase in flux at best. Interestingly, because Jupiter is composed in large part of degenerate hydrogen fluid, an increase in mass leads to very little increase in radius and, hence, reflecting area. Thus, it is difficult to make a Jupiter detectable using HST. For the initially conceived 0.85 m aperture SIRTF mirror, at 20 μm the first dark Airy ring is at nearly 6 arcsec. This is too far to permit of separation of the Jupiter image from that of the parent star (COMPLEX, 1990). However, in this case much of the planetary emission is thermal, so that one might expect to be able to detect a Jupiter in a much more distant orbit around its parent star using SIRTF. Nonetheless, these examples illustrate the difficulty of direct detection using existing or planned large spaceborne telescopes.

Ground-based telescopes have the advantages of larger surface area, use of sidelobe suppression techniques, and more sensitive detectors. However, resolution is seeing-limited (owing to atmospheric turbulence) rather than diffraction-limited. Use of adaptive optics may potentially overcome this problem (Angel, 1994). Use of interferometers for direct detection also may afford significant advantages over single telescopes in the near future. Extrapolation of such techniques leads to the possibility of doing spectroscopy on detected planets to search for disequilibrium atmospheric species (such as ozone, derived from molecular oxygen) as a proxy indicator for life. Such prospects are exciting, but remain hypothetical until the capability for direct detection of planets is achieved and yields positive results.

Direct detection is, of course, the technique of choice for protoplanetary systems. As discussed in the previous section, investigation of high mass gas disks, later-stage debris disks, bipolar flows and other phenomena associated with active star formation continues to bear fruit using a range of telescopes from optical through millimeter wavelengths. Further advances in adaptive optics and interferometry will yield higher spatial resolution. Observations of disks with spatial resolution of 1–10 AU is an achievable goal in the coming decade, and will allow of testing of hydrodynamical models of disk evolution processes related to planet formation. Nonetheless, as discussed above, study of disks does not directly inform us as to the frequency of mature planetary systems. To survey nearby galactic space with this issue in mind, in view of the limitations of direct detection, indirect techniques have been employed for decades.

Indirect Searches

Indirect searches involve detecting some signature of a planetary system in the motion or brightness of the central star. The two classical techniques are astrometry and radial velocity. Astrometry involves determining the sky-plane component of a star's motion associated with the barycentric wobble induced by planets; this motion is an ellipse of angular semimajor axis

$$x = \frac{am}{Mr} \tag{21.2}$$

where m is the mass of the planet, M the stellar mass, r the observer's distance to the star and a the physical semimajor axis, which is related to the period P by

$$a = (M + m)^{1/3}P^{2/3} \tag{21.3}$$

The challenge in observing the wobble is clear: for an observer 10 pc from the solar system, Jupiter induces a wobble amplitude of half a milliarcsecond (mas). This compares with a typical proper motion of the system, if the observer–Sun relative velocity is 10 km s^{-1}, of 0.2 arcsec yr^{-1} (COMPLEX, 1990). The other difficulty is time: one must detect the wobble over much of an orbital period, and more realistically multiple periods, to be sure of a detection. This can be further complicated by the effects of multiple planets, but for our solar system the Jupiter signal would clearly dominate. One active search program uses a special Ronchi ruling to create an accurate metric, and has an average accuracy per night of somewhat better than 3 mas (Gatewood, 1987), which is a vast improvement over traditional searches using photographic plates, which could achieve at best 100 mas (COMPLEX, 1990).

Radial velocity observations rely on determining changes in the radial component of the star's motion due to orbiting planets by measuring the periodic Doppler shift of lines in the star's atmosphere,

$$\frac{\Delta\lambda}{\lambda} = \frac{v}{c}\sin i \tag{21.4}$$

where λ and $\Delta\lambda$ are the wavelength and wavelength shift, v the orbital speed of the star, c the speed of light, and i the angle between the perpendicular to the plane of the orbit and the line of sight.

The velocity and planetary mass are related by

$$v = m \sqrt{\frac{G}{Ma}} \tag{21.5}$$

where G is the gravitational constant and a again can be found by using eq. (21.3) with the observationally determined period (COMPLEX, 1990). A serious ambiguity associated with this technique lies in its inability to determine i, which in turn prevents one from solving unambiguously for the mass. For randomly oriented orbits, $\sin i = 0.87$ (assuming a typical $i = 60°$), but the utility of such a number is questionable. For a large ensemble of detections, such statistical considerations could be used to determine what fraction of such detections are likely to be stellar mass bodies 'masquerading' as planets; but for a single detection one cannot eliminate the possibility that $\sin i$ is close to zero and one's prospective 'planet' is actually a cool star or brown dwarf.

Current techniques can achieve velocity precision of $\pm 5 \, \mathrm{m \, s^{-1}}$ but a key challenge with the approach is the need to eliminate other natural sources of Doppler shift in the stellar photospheric lines. For Jupiter orbiting the Sun, seen from 10 pc distance, the amplitude of the velocity perturbation is roughly $15 \, \mathrm{m \, s^{-1}}$ (COMPLEX, 1990). Stellar rotation rates are typically $1 \, \mathrm{km \, s^{-1}}$; the resulting Doppler shift can be fairly readily removed, though one has to worry about variable rotation rates versus latitude on the stellar disk. Much more serious is the Doppler shift caused by turbulent processes on the photospheric disk. Typical convective motions yield a systematic blueshift in absorption lines; when modulated by stellar cycles analogous to the Sun's 11 year solar cycle such shifts could be mistaken for the signature of a planet. Initial studies directly observing solar CO lines (Deming et al., 1987) suggested that the amplitude of the mimicked planet perturbation could well exceed several tens of meters per second. However, more recent studies dispute this. McMillan et al. (1993) used the Moon as a solar reflector, allowing observing techniques and equipment much more analogous to those used in stellar radial velocity searches. Using the steep flanks of solar violet lines, they find amplitudes less than $10 \, \mathrm{m \, s^{-1}}$, and hence conclude that such intrinsic photospheric variations would not overwhelm the signature of a Jupiter-type planet around a nearby star.

Other 'special' effects such as stellar flares and variable stars must also, of course, be of concern; systematic measurements in several lines as well as simultaneous photometry of the star would appear to be essential to correct for such possibilities. While no confirmed detection of Jupiter-mass or smaller bodies have occurred using the radial velocity technique (Campbell et al., 1988; McMillan *et al.*, 1993), the high precision makes feasible the ability to map out the incidence of planetary systems in nearby space (~10 pc) in the coming decade, provided that enough observing time on telescopes of moderate size is available.

Figure 21.2 (COMPLEX, 1990) illustrates that radial velocity and astrometric techniques are complementary to each other. Plotted is a search space comprised of planetary mass (relative to the primary star) versus semimajor axis, for a system 10 pc from the observer. For astrometry two assumed accuracies are given, 10^{-3} and 10^{-5} arcsec, and for radial velocity an accuracy of 10 m s^{-1} is assumed. Two parent star masses, 0.3 and 1 solar masses, are used. Areas above the corresponding lines represent detectable planet mass and semimajor axis combinations. Note that radial velocity and astrometry are most sensitive for small and large semimajor axes, respectively. Note also that with current technologies, none of the planets in our solar system would be detectable, but improvements by a factor of 10 in astrometry would open up detection of Jupiter- and Saturn-type objects. Doppler techniques today are well able to handle Jupiter-mass planets at semimajor axes

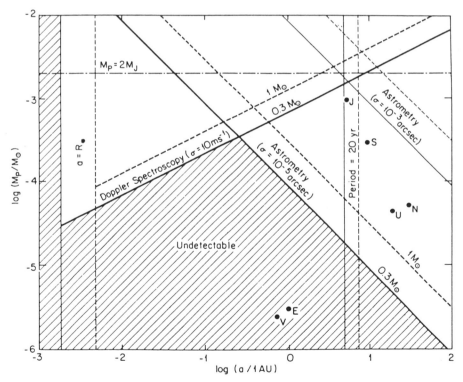

Figure 21.2 Discovery space for planets using astrometry and radial velocity techniques, from COMPLEX (1990). See text for details.

of an AU or so. The vertical lines at 20 years provide an indication of limits imposed by human patience.

A somewhat different application of radial velocity techniques has led to the discovery of two planet-mass objects orbiting a millisecond pulsar PSR 1257+12 (Wolszczan & Frail, 1992). Observations using the Arecibo radio telescope revealed two periodicities in the period of the pulsar which correspond to objects with masses (in units of Earth masses) $3.4/\sin i$ and $2.8/\sin i$ in near-circular orbits at distances from the neutron star of 0.36 and 0.47 AU, respectively. The detection of objects in this type of system relies on the high rotational stability of millisecond pulsars. While precession and nutation of the pulsar can also lead to modulation of the period of its radio signal, and possibly cannot be ruled out for PSR1257+12 (Gil & Jessner, 1993), recent support for the planet interpretation has come in the form of observed modulation of the signal exactly as predicted by mutual gravitational perturbation of the two planets (Wolszczan, 1994).

Other indirect techniques for detecting planetary systems involve photometric observation of stars to detect brightness changes. If planetary systems are common, some may be oriented such that transits of the planets across the stellar face would lead to temporary and periodic drops in brightness. These changes are small, less than a part in 10 000 for Earth-like planets around a solar-type star, but could be distinguished from other intensity fluctuations by observing color changes due to differential limb darkening as the planet crosses the star's disk (Burke, 1992). Transit dimming will be most obvious for giant planets in close orbit around M-dwarfs, but in any event accurate systematic searches using this technique would need to be done from space, perhaps with large-aperture CCDs to simultaneously keep track of as many stars as possible (Burke, 1992).

A novel approach to serendipitous discovery of planets is to observe gravitational lensing effects in stars. As stars move relative to each other through the galactic disk, the chance passage of one into the line of sight between an observer and target star leads to lensing events lasting typically a month. Single stars acting as lenses yield bell-shaped curves; multiple systems (including stars with planets) as lenses lead to overlapping curves at microarcsecond spacing, leading to a non-bell-shaped curve ('microlensing': Mao & Paczynski, 1991). Gould & Loeb (1992) point out that if Jupiters form preferentially at the water–ice condensation zone (because of the large amount of solids available for core formation: Stevenson & Lunine, 1988), then such systems, if halfway to the galactic center as seen from Earth, have preferential geometries for microlensing events. Searches for events toward the galactic center (a logical direction because of the rich field of stars) would have an enhanced probability of detecting planets by this argument, a factor of 5–10 over an assumed random orbital spacing of planets from their parent star. Recent reported detections of microlensing events of large magellanic cloud giant stars by old stars or brown dwarfs in our galaxy (Alcock *et al.*, 1993; Aubourg *et al.*, 1993; Udalski *et al.*, 1993) enhance the attractiveness of this technique. The prospect of a surprise discovery of a planet by this technique is intriguing, though the small probability for lensing a galactic bulge star makes it less useful as a tool for statistically determining the incidence of planetary systems.

Are Planetary Systems Common?

Although no planetary system has been detected around a main sequence star other than our Sun, vast observational and theoretical progress has been made in the last ten years. Based on this, the following rather general statements are offered:

1. For the most abundant stars in the Galaxy, those of solar mass and less, formation of a gas-dust disk is very common; perhaps half of such stars possess disks during their pre-main-sequence phase.
2. The fraction of disks which form planets is uncertain, in part because close binary systems may also be products of disks. A reasonable guess, not contradicted by observation or hydrodynamical modeling of such disks, suggests that perhaps half of such disks manage to evolve through the high-mass unstable (binary-forming) phase, and, hence, are capable of making planets.
3. The presence of planets around one pulsar is best understood as the result of disk formation as a consequence of the supernova explosion which formed the neutron star. The existence of this system therefore suggests that planet formation is a reasonably robust process which can take place in at least two very different physical situations.
4. Searches for planetary systems around main sequence stars are limited in the near future, by measurement accuracies, to direct or indirect detection of Jupiter-class planets (gas giants) around stars in the solar neighborhood. It is as yet unclear whether giant planet formation always accompanies formation of smaller, rock-ice terrestrial-type planets. However, the conditions for giant planet formation appear on theoretical grounds to be more constrained than those for terrestrial planets; hence, negative search results using current technologies cannot rule out the presence of purely terrestrial-planet-type (or Uranus–Neptune-type) systems.
5. Current search techniques and instrumentation have not yet led to searches of sufficient sensitivity and completeness to enable any meaningful statement to be made about the frequency of planetary systems. Searches envisioned in the near-term using ground-based and Earth-orbital facilities could do so over a two-decade timespan. Longer-term searches, including those taking advantage of a postulated lunar infrastructure, could over several decades provide meaningful information on the frequency of Earth-mass planets.

Planetary systems may well be common throughout our Galaxy; there are no data which contradict, for example, the assignment of $f_p > 0.1$ in eq. (21.1), and some data mitigate in its favor. However, no inquiring human mind would be satisfied leaving the problem at that. The excitement inherent in the question of the incidence of planetary systems lies in the fact that we now have technology capable of addressing this question, while we remain as ignorant of the answer as all the rest of humankind before us. For the adventurous soul, this is a great place to be.

Acknowledgements

I want to thank Ben Zuckerman, Gene Levy and Bill Hubbard for valuable discussions and source material. The chapter is dedicated to the late Dr Robert Harrington. Bob's enthusiasm for the search, particularly evident to me at a 1985 conference on brown dwarfs at George Mason University, helped expand my interests outward from our own solar system. Preparation of this work was supported by NASA's Origins of Solar Systems Research Program.

References

ADAMS, F. C. & LIN, D. N. C. (1993). In Levy and Lunine, *op. cit.*, pp. 721–48.

ANGEL, J. R. P. (1994). *Nature*, **368**, 203–7.

ALCOCK, C. *et al.* (1993). *Nature*, **365**, 621–3.

AUBOURG, E. *et al.* (1993). *Nature*, **365**, 623–5.

BACKMAN, D. E. & PARESCE, F. (1993). In Levy and Lunine, *op. cit.*, pp. 1253–304.

BASRI, G. & BERTOUT, C. (1993). In Levy and Lunine, *op. cit.*, pp. 543–66.

BECKWITH, S. V. W. & SARGENT, A. I. (1993). In Levy and Lunine, *op. cit.*, pp. 521–41.

BLACK, D. C. (1986). In *Astrophysics of Brown Dwarfs*, ed. M. C. Kafatos, R. S. Harrington & S. P. Maran, pp. 139–47. Cambridge: Cambridge University Press.

BLACK, D. C. & MATTHEWS, M. S., (eds.) (1985). *Protostars and Planets II*. Tucson: University of Arizona Press.

BODENHEIMER, P. (1993). In Phillips *et al.*, *op. cit.*, pp. 257–64.

BODENHEIMER, P., RUZMAIKINA, T. & MATHIEU, R. D. (1993). In Levy and Lunine, *op. cit.*, pp. 367–404.

BURKE, B. F., (ed.) (1992). *TOPS: Toward Other Planetary Systems*. Washington, DC: Solar System Exploration Division, NASA.

CAMPBELL, B., WALKER, G. A. H. & YANG, S. (1988). *Astrophys. J.*, **331**, 902–21.

COMPLEX (Committee on Planetary and Lunar Exploration) (1990). *Strategy for the Detection and Study of Other Planetary Systems and Extrasolar Planetary Materials: 1990–2000*. Washington, DC: National Academy Press.

DEMING, D., ESPENAK, F., JENNINGS, D., BRAULT, J. & WAGNER, J. (1987). *Astrophys. J.*, **316**, 771–87.

DINER, D. J. & APPLEBY, J. F. (1986). *Nature*, **322**, 436–8.

DOLE, S. H. (1970). *Icarus*, **13**, 494–508.

DUNCAN, M., QUINN, T. & TREMAINE, S. (1988). *Astrophys. J. Lett.*, **328**, L69–L73.

GATEWOOD, G. D. (1987). *Astron. J.*, **97**, 1189–96.

GIL, J. A. & JESSNER, A. (1993). In Phillips *et al.*, *op. cit.*, pp. 71–9.

GOLDREICH, P. & WARD, W. R. (1973). *Astrophys. J.*, **183**, 1051–61.

GOULD, A. & LOEB, A. (1992). *Astrophys. J.*, **396**, 104–14.

HUBBARD, W. B. (1977). *Icarus*, **30**, 305–10.

ISAACMAN, R. & SAGAN, C. (1977). *Icarus*, **31**, 510–33.

LAPLACE, P. S. DE (1796). *Exposition du système du monde* (Paris).

LEVY, E. H., (1993). In Phillips *et al.*, *op. cit.*, pp. 181–95.

LEVY, E. H. & LUNINE, J. I., (eds). (1993). *Protostars and Planets III*. Tucson: University of Arizona Press.

LIN, D. N. C., WOOSLEY, S. E. & BODENHEIMER, P. (1991). *Nature*, **353**, 827–9.

LISSAUER, J. J. & STEWART, G. R. (1993). In Levy and Lunine, *op. cit.*, pp. 1061–88.

McMILLAN, R. S., MOORE, T. L., PERRY, M. L. & SMITH, P. H. (1993). *Astrophys. J.*, **403**, 801–9.

MAO, S. & PACZYNSKI, B. (1991). *Astrophys. J. Lett.*, **374**, L37–L40.

MUMMA, M. J., WEISSMAN, P. R. & STERN, S. A. (1993). In Levy and Lunine, *op. cit.*, pp. 1177–252.

PHILLIPS, J. A., THORSETT, S. E. & KULKARNI, S. R., (eds.) (1993). *Planets around Pulsars*. San Francisco: Astronomical Society of the Pacific Conference Series Vol. 36.

PODOLAK, M., HUBBARD, W. B. & POLLACK, J. B. (1993). In Levy and Lunine, *op. cit.* pp. 1109–47.

PODSIADLOWSKI, P., PRINGLE, J. E. & REES, M. J. (1991). *Nature*, **352**, 783–4.

SAUMON, D., HUBBARD, W. B., CHABRIER, G. & VAN HORN, H. M. (1992). *Astrophys. J.*, **391**, 827–31.

SHKLOVSKII, I. S. & SAGAN, C. (1966). *Intelligent Life in the Universe*. San Francisco: Holden-Day.

SMITH, B. A. & TERRILE, R. J. (1984). *Science*, **226**, 1421–24.

SMITH, E. V. P. & JACOBS, K. C. (1973). *Introductory Astronomy and Astrophysics*. Philadelphia: W. B. Saunders.

STAHLER, S. W. & WALTER, F. M. (1993). In Levy and Lunine, *op. cit.*, pp. 405–28.

STEVENSON, D. J. (1984). *Lunar Planet. Sci. Conf. XV*, pp. 822–3 (abstract).

STEVENSON, D. J. & LUNINE, J. I. (1988). *Icarus*, **75**, 146–55.

STROM, S. E., EDWARDS, S. & SKRUTSKIE, M. F. (1993). In Levy and Lunine, *op. cit.*, pp. 837–66.

STRUCK-MARCELL, C. (1988). *Astrophys. J.*, **330**, 986–91.

UDALSKI, A. *et. al.* (1993). *Acta Astronom.*, **43**, 289.

WEIDENSCHILLING, S. J. & CUZZI, J. N. (1993). In Levy and Lunine, *op. cit.*, pp. 1031–60.

WEISSMAN, P. R. (1990). *Nature*, **344**, 825–30.

WETHERILL, G. W. (1993). *Lunar Planet. Sci. Conf. XXIV*, pp. 1511–12 (abstract).

WOLSZCZAN, A. & FRAIL, D. A. (1992). *Nature*, **355**, 145–7.

WOLSZCZAN, A. (1994). *Science*, **264**, 538–42.

WOOD, J. A. & MORFILL, G. E. (1988). In *Meteorites and The Early Solar System*, ed. J. F. Kerridge & M. S. Matthews, pp. 329–47. Tucson: University of Arizona Press.

WUCHTERL, G. (1990). *Icarus*, **91**, 53–64.

ZHARKOV, V. N. & GUDKOVA, T. V. (1991). *Ann. Geophys.* **9**, 357–66.

ZINNECKER, H., MCCAUGHREAN, M. J. & WILKING, B. A. (1993). In Levy and Lunine, *op. cit.*, pp. 429–95.

ZUCKERMAN, B. (1993). In Phillips *et al.*, *op. cit.*, pp. 303–15.

ZUCKERMAN, B. & BECKLIN, E. E. (1993). *Astrophys. J.*, **406**, L25–L28.

22

Atmospheric Evolution, the Drake Equation and DNA: Sparse Life in an Infinite Universe

MICHAEL H. HART

Atmospheric Evolution

During the 4 ½ billion years of geologic time, the atmosphere of the Earth has changed markedly. Of the many processes which have played a role in that evolution, the most important are listed in table 22.1.

Table 22.1. *Important processes in atmospheric evolution*

1. Degassing of volatiles from interior
2. Condensation of water vapor into oceans
3. Solution of CO_2 and other gases in seawater
4. Fixing of CO_2 in carbonate minerals (Urey reaction)
5. Photodissociation of water vapor
6. Escape of hydrogen into space
7. Development of life, and variations in the biomass
8. Net photosynthesis and burial of organic sediments
9. Chemical reactions between atmospheric gases
10. Oxidation of surface minerals
11. Changes in solar luminosity
12. Changes in albedo (cloud cover, ice cover, etc.)
13. The greenhouse effect

There are good reasons (Brown, 1952) to believe that at some early stage the Earth had almost no atmosphere, having lost whatever gaseous envelope, if any, that it started with. Our present atmosphere is derived from materials degassed

215

from the interior of the Earth; perhaps largely from volcanoes, but also from fumaroles, and by a slow seepage through the crust.

Among the gases released was water vapor, most of which eventually condensed to form the oceans. The history of the other gases is highly complex. Some molecules were broken apart in the upper atmosphere by the action of sunlight. The lightest gases (hydrogen, helium) were able to escape into outer space (Spitzer, 1952). Other gases, such as carbon dioxide, are highly soluble in seawater and were able to react chemically with minerals dissolved in the oceans (Urey, 1951, 1952).

The origin of life, and subsequent biochemical processes (such as photosynthesis), have powerfully affected the composition of the atmosphere, as have various inorganic reactions such as the oxidation of surface minerals. Meanwhile, there have been marked changes in the Earth's surface temperature, caused in part by variations in the Sun's luminosity, by variations in the Earth's reflectivity or albedo, and by the greenhouse effect.

In principle, if one knew the exact rate at which all these processes occurred, one could trace on a high-speed computer the entire evolution of the Earth's atmosphere over the past 4 ½ billion years. In an earlier paper (Hart, 1978), I have described in some detail the formulas which were used and the approximations which were made in constructing such a computer simulation. In view of the various approximations and uncertainties involved, one cannot expect the results of such a computer simulation to be reliable in every detail, but they are consistent with the available observational data, and they probably indicate the general pattern of our atmosphere's evolution fairly well.

According to the computer simulation, the Earth was probably a good deal warmer during the first 2.5 billion years of geologic time than it is today. It cooled down to about its present temperature roughly 2 billion years ago, when free oxygen first appeared in the atmosphere, and when various other gases capable of causing a large greenhouse effect were largely eliminated by oxidation.

Since the early Earth seems to have been quite warm, it is natural to wonder how much hotter it might have been if the Earth were situated somewhat closer to the Sun. It is fairly easy to modify the computer program so as to simulate the effect of a smaller Earth–Sun distance. The results are quite striking: If the Earth's orbit were only 5% smaller than it actually is, during the early stages of Earth's history there would have been a 'runaway greenhouse effect', and temperatures would have gone up until the oceans boiled away entirely!

This result was not entirely unexpected. A similar conclusion (although based on a less detailed model) had been reached previously by Rasool & de Bergh (1970); and it is widely believed that a runaway greenhouse effect actually occurred on Venus, which is 28% closer to the Sun than the Earth is. More surprising, perhaps, were the results of computer runs which simulated the effect of a some-what larger Earth–Sun distance. Those runs indicate that if the Earth–Sun distance were as little as 1% larger, there would have been runaway glaciation on Earth about 2 billion years ago. The Earth's oceans would have frozen over entirely, and would have remained so ever since, with a mean global temperature of less than −50 °F. (Similar conclusions, although derived from quite different models, were reached earlier by Budyko, 1969, and by Sellers, 1969.) Taken together, these computer runs indicate that the habitable zone about our Sun is not wide, as Huang (1959, 1960) had suggested, but is instead quite narrow.

What about the habitable zones about other stars? How large are they? Here, too, it is possible to modify the original computer program to simulate the effect of the more intense radiation from a larger star, or the weaker radiation from a smaller one. The modifications needed are a bit tricky, since large stars evolve more rapidly than small stars, and their relative luminosities change with time. But the required modifications can be made, and again the results are quite striking. The computer simulations (Hart, 1979) indicate that a star whose mass, M_{star}, is less than $0.83\ M_{Sun}$ will have no zone about it which is continuously habitable. If a planet is far enough from such a star to avoid a runaway greenhouse effect in its early years, it will inevitably undergo runaway glaciation somewhat later in its history. Nor, according to the computer results, does a star heavier than 1.2 solar masses have any continuously habitable zone about it.

Similar calculations indicate that the size of the planet itself has a profound effect on the evolution of its atmosphere. Unless a planet has a mass within the range $0.85\ M_{Earth} < M_{planet} < 1.33\ M_{Earth}$, it cannot – regardless of its distance from the Sun – maintain moderate surface conditions for more than 2 billion years.

Calculation of N, Using a Modified Drake Equation

The galaxy we are in, the Milky Way Galaxy, contains upwards of 100 billion stars, many of which appear to be quite similar to our Sun, and many of which may have planets orbiting about them. Within range of our large telescopes there are at least 10^9 other galaxies – possibly 10^{11} or more – together totalling at least 10^{20} stars.

In view of this enormous number of stars, it is quite natural to ask two questions: Of this vast multitude of stars, how many have planets near them which are suitable for the evolution of life? And on how many of those planets has life actually arisen? More intriguing still – since we are naturally more interested in *intelligent* life – are the questions: How many advanced civilizations can we expect to exist in the Milky Way Galaxy? How many can we expect in the entire universe?

To estimate N, the expected number of advanced civilizations in a typical galaxy the size of the Milky Way, many writers use as a starting point some version or modification of the well-known Drake equation. The version which I shall use is:

$$N = N_{Gal} \cdot f_{popl} \cdot f_{PMR} \cdot f_{PS} \cdot f_{HP} \cdot f_{life} \cdot f_{intel} \cdot f_{tech} \qquad (22.1)$$

In this equation, N_{Gal} denotes the total number of stars in the Galaxy; f_{popl} represents the fraction of those which are population type I stars; f_{PMR} denotes the fraction of population type I stars which are within the 'proper' mass range (i.e. stars which are neither too large nor too small to have continuously habitable zones about them); f_{PS} represents the fraction of such stars which have planetary systems; f_{HP} denotes the fraction of planetary systems which include a habitable planet (i.e. a planet whose size, composition and distance from its sun make it suitable for the development of life); f_{life} represents the fraction of habitable planets upon which life actually arises; f_{intel} is the fraction of those planets on which intelligent life forms (i.e. $\geq$ human intelligence) eventually evolve; and f_{tech} denotes the fraction of those which develop and sustain advanced technologies (avoiding destruction by nuclear war, plagues, ecological disasters, etc.)

N_{Gal} is usually estimated to be about 2×10^{11}. Of those, about 50% should probably be excluded because the gases from which they condensed had too low an abundance of heavy elements for large, solid planets like the Earth to be formed. According to the calculations described above (p. 217), the 'proper' mass range is $0.83\,M_{Sun} < M_{star} < 1.2\,M_{Sun}$. Direct star counts in the solar neighborhood indicate that about 10% of stars fall within that range.

The value of f_{PS} is still in doubt; suppose we estimate it to be about 10%. (As 50% or more of stars are members of double or multiple star systems, that can hardly be much of an underestimate.)

If the fairly involved calculations described on pp. 216–217 are correct, only about one planetary system in a hundred (even if the central star is in the proper mass range) contains a habitable planet. That would make $f_{HP} \sim 10^{-2}$. The value of f_{life} is extremely speculative; for the moment, let us defer trying to estimate it. However, if life ever does arise on a planet, the process of darwinian evolution should frequently lead to advanced life forms, and f_{intel} may well be as high as 10%. For the final factor, f_{tech}, a guess of 50% might be in order.

Since the value of f_{life} is so speculative, we might combine all the other factors in eq. (22.1) together, and write it as

$$N = N_{combo} \cdot f_{life}. \qquad (22.2)$$

If we combine the various numerical estimates given above, we obtain the result $N_{combo} \sim 5 \times 10^5$. However, as all the factors which go into N_{combo} are highly uncertain, its true value could be very different. Some optimists have estimated N_{combo} to be as high as 10^9, or perhaps even a bit larger; while if very conservative estimates are used for the various factors, N_{combo} could be only 10^1, or even less.

Now if f_{life} has a very low value – for example, 10^{-15} – this uncertainty in N_{combo} is unimportant. For in that case, any plausible value of N_{combo} results in $N \ll 1$. However, if f_{life} has a moderate value – say 10^{-2} – then the uncertainty in N_{combo} renders the 'Drake equation approach' virtually useless as a method of deciding whether advanced civilizations are frequent in a typical galaxy, or whether the majority of galaxies do not contain even a single civilization. Nor, given the highly speculative nature of factors such as f_{intel} or f_{tech}, can we expect to obtain a reliable estimate of N_{combo} within the foreseeable future. What method, then, could we use to estimate the value of N?

Our Failure To Observe Extraterrestrials

I would suggest that in that case the best way to approach the problem of estimating N would not be by examining the factors which *cause* N to have a certain value, but rather by taking an empirical approach and considering the various *effects* which we might expect to observe if N had a given value.

If, for example, there were 100 000 advanced civilizations scattered about the Milky Way Galaxy, what observable effects might we expect to see? Well, if there were really so many technologically advanced races in our Galaxy, then surely at least *one* of them would have explored and colonized the Galaxy, just as we humans have explored and colonized this planet. Various estimates (Hart, 1975; Jones, 1976; Papagiannis, 1978) indicate that no more than a few million years would be needed to colonize most of the Galaxy. Since that is very much less than the age of our Galaxy ($\geqslant 10^{10}$ years), if N were really as large as 100 000 then the solar system would have been colonized by extraterrestrials a long time ago, and we would see them here today.

But, of course, we do not see any extraterrestrials, either on Earth or anywhere else in the solar system. There is no indication that the solar system was ever visited by extraterrestrials; and, quite obviously, we have not been colonized. It can reasonably be concluded, therefore, that N is *not* equal to 100 000. The same argument, of course, would rule out any other large number. It would not, though, completely rule out the possibility that there were a *small* number of civilizations in our Galaxy, none of which were interested in interstellar exploration and colonization (nor ever had been, in all the ages since they first acquired the technological capability).

N, therefore, is a small number, possibly a very small number; and our conclusion, since it has an empirical basis (i.e. the absence of extraterrestrials on Earth) cannot be upset by any unreliable calculations based on the Drake equation. Nevertheless, it would certainly be interesting to know just *how* low N is. I would like to suggest that a realistic calculation of f_{life} indicates that it is an extremely low number, and that N therefore is also extremely low.

Calculation of f_{life}

Before attempting to compute f_{life}, we should perhaps first explain what we mean by the word 'life'. It is difficult to give an exact definition of this term (see Feinberg and Shapiro, 1980, for an interesting and novel approach), but we might roughly say that a living organism is an object which feeds and reproduces. (An object 'feeds' if it ingests and chemically transforms material in its environment into chemicals which it is itself composed of.)

The living organisms which we see on Earth are all composed of complex carbon compounds in an aqueous medium. A wide variety of such compounds are found in most organisms, but the two most significant types are: (1) the proteins, which are large, complex molecules consisting of long strings of simpler components called amino acids; and (2) the nucleic acids, which consist of long strings of simpler components called nucleotides. (The most important type of nucleic acid, DNA, contains four different nucleotides, each occurring many times in a single molecule of DNA.) The proteins perform a crucial role in catalyzing essential biochemical reactions, while the nucleic acids perform an even more vital role by storing the hereditary information which allows organisms to reproduce, and by directing the synthesis of proteins. Nucleic acids are the primary genetic material, and they contain (in coded form) instructions for synthesizing the organism and its components. The code is based on the number of each of the four types of nucleotides in a given strand, and on the *order* in which those different nucleotides are arranged.

Now if there is life on other planets in the universe, it is perfectly possible that the organisms on such planets use quite different compounds to perform the tasks which in terrestrial organisms are carried out by the proteins and the nucleic acids. But since those tasks are so difficult, detailed and varied, the compounds carrying them out would of necessity have to be just about as large and as complex in structure as are the proteins and nucleic acid molecules which we find in terrestrial organisms.

How large, then, is f_{life}, which is defined as the probability that life will actually arise on a given planet which has a wholly suitable environment. We may safely assume that on such a planet the surface temperatures are suitably moderate, that liquid water is present in ample quantity, and that simple compounds of carbon, oxygen, hydrogen and nitrogen are abundant. Many experiments (see Miller & Orgel, 1974, for a partial list) have shown that a combination of such chemicals will, in the presence of electric discharges, react to produce a variety of more complex organic molecules, including amino acids. Under suitable conditions, short chains of amino acids have also been produced.

This is an encouraging start. However, in order to have living organisms, some sort of genetic material – such as DNA – must be present also. Experiments simulating primitive Earth conditions have not, to date, resulted in the formation of nucleotides; but simpler compounds related to them have been produced in such experiments, and it is not unduly optimistic to assume that nucleotide molecules too will naturally be formed on a suitable planet.

To induce those molecules to polymerize into nucleic acid strands (under the assumed primitive Earth conditions) is a bit of a problem, but not a hopeless one. It is, though, crucial for the proper functioning of the resulting nucleic acid molecule that the various nucleotides in the strand are arranged in the correct order. Two different nucleic acid strands, even if of exactly the same length, will not normally be biologically equivalent unless they contain the same nucleotide bases arranged in the same order.

The great majority of possible nucleic acid molecules are quite useless (or even harmful) biologically. Most of the others are useful only in an organism which already has many other genes. Let us suppose, however, that there exists a particular DNA molecule – 'genesis DNA' – which, if introduced into some primitive conglomeration of proteins, lipids, nucleotides and their building blocks, will both replicate properly and perform some useful biological function. In other words, we are supposing that the formation of a single molecule of genesis DNA and its introduction into a suitable environment will suffice to create a viable organism and to get the process of darwinian evolution started.

To simplify our calculations, let me make a few more assumptions (admittedly, rather optimistic ones).

(1) Under the conditions prevailing on a primitive Earth-like planet, not just amino acids but also nucleotides will be readily formed.
(2) Those same conditions will favor the polymerization of nucleotides.
(3) Uniform helicity of the resulting strands is thermodynamically favored.
(4) A strand of genesis DNA is quite short, as genes go, containing only 600 nucleotide residues.
(5) There exists some chemical effect which favors the formation of nucleic acid strands of that length.

If these assumptions are valid, then a large number of strands of nucleic acid, each consisting of about 600 nucleotides, will be formed spontaneously on any suitable planet. Let us calculate the probability that one of those strands will have its bases arranged in the right order, i.e. in the same order as in genesis DNA.

There are only about 2×10^{44} nitrogen atoms near the surface of the Earth, or in its atmosphere. As a single 600 residue strand of nucleic acid includes more than 2000 nitrogen atoms, there could have been no more than about 10^{41} strands of DNA existing together on the primitive Earth at any given moment. If every such strand could split up and recombine with other fragments at a rate of 30 times a second, then in 1 year (roughly 3×10^7 s) a maximum of 10^{50} different strands could be formed, and in 10 billion years a maximum of 10^{60}. (This, obviously, is a strong maximum.)

The number of conceivable arrangements of the four different nucleotides into a strand of DNA 600 residues long is 4^{600}, which is about 10^{360}. The chance that a *particular* one of them would be formed spontaneously – even in 10 billion years – is therefore extremely small, 10^{-300}. However, the chance of forming genesis DNA is not necessarily that low. It has been demonstrated that at some positions in a strand of nucleic acid it is possible to replace one nucleotide base by another without changing the biological effect of the strand. Let us suppose (very optimistically) that in a strand of genesis DNA there are no fewer than 400 positions where any one of the four nucleotide residues will do, and at each of 100 other positions either of two different nucleotides will be equally effective, leaving only 100 positions which must be filled by exactly the right nucleotides. This appears to be an unrealistically optimistic set of assumptions; but even so, the probability that an arbitrarily chosen strand of nucleic acid could function as genesis DNA is only one in 10^{90}. Even in 10 billion years, the chance of forming such a strand spontaneously would be only $10^{-90} \times 10^{60}$, or 10^{-30}.

There are several reasons why the true value of f_{life} is very much lower than 10^{-30}. In the first place, we have ignored all the difficulties involved in producing nucleotides abiotically, in concentrating them in a small region, of preventing their spontaneous destruction and in getting them to polymerize in an aqueous environment. In the second place, a DNA molecule cannot direct protein synthesis unless certain other complex organic molecules (called 'transfer RNA') are present; nor can it even replicate itself spontaneously in the absence of certain other organic catalysts (DNA polymerases). Unless these other compounds had already been formed (how?) and were in the immediate vicinity, even if a molecule of genesis DNA happened to be formed, it would be unable to function. And in the third place, the assumption that there exists a gene – genesis DNA – which, without any other genes present, can produce a viable organism is highly optimistic. The simplest known organism which is capable of independent existence includes about 100 different genes. For each of 100 different specific genes to be formed spontaneously (in 10 billion years) the probability is $(10^{-30})^{100} = 10^{-3000}$. For them to be formed at the same time, and in close proximity, the probability is very much lower.

Probability and Selection

The conclusion reached above, that the probability of life arising on a given planet – no matter how favorable conditions on that planet might be – is less than one in 10^{30}, is perhaps somewhat surprising. If f_{life} is so low, you may ask, what are *we* doing here?

This leads to an interesting philosophical question: If we calculate the probability of an event occurring to be a very low number, and the event then occurs, does it show that our calculation is wrong? For example, a simple calculation shows that the probability of tossing an honest coin 40 times and getting 40 consecutive heads is $\frac{1}{2}^{40}$, or about one in a trillion. Suppose, though, that you took a particular coin, flipped it 40 times and got 40 heads. Would you then rush about excitedly, telling everyone about the 'almost unbelievable' coincidence which had occurred, and send in a report to a scientific journal? Of course not! You would simply conclude that the coin was not balanced, and that your calculations therefore did not apply. Suppose, however, that you made not just one set of 40 flips, but 10^{12} different runs, each of 40 flips. And further suppose that one of those runs resulted in 40 consecutive heads. In that case you would conclude that the coin was honest, that your calculations were correct and that no unbelievable coincidence had occurred. Similarly, if we were shown an (undoctored) film displaying a run of 40 consecutive heads, we would normally interpret it not as evidence of a remarkable coincidence, but merely as evidence that the coin was unbalanced. However, if we knew that the maker of the film had made and photographed 10^{12} runs, each of 40 flips, but only let us see the film of the one successful run, we would see no reason to doubt the correctness of our calculations.

Life in the Infinite Universe

Why do I suggest such a fanciful possibility? Because the universe we live in is not finite, but infinite! Modern astronomical observations strongly support the so-called 'big bang' cosmology, and the majority of the evidence indicates that our universe is open and will continue to expand indefinitely (Gott *et al.*, 1974). Analysis shows that, unless a very unusual topology is assumed, such an open universe must be infinite in extent, with an infinite number of galaxies, an infinite number of stars and an infinite number of planets. In an infinite universe, any event which has a finite probability – no matter how small – of occurring on a single given planet must inevitably occur on some planet. In fact, such an event must occur on an infinite number of planets. (See Ellis & Brundrit, 1979, for an interesting discussion of this point.)

We are therefore in the position of the hypothetical film-viewer described above. There are an infinite number of habitable planets in the universe. On each of these, nature patiently tosses her tetrahedral dice for 10 billion years, trying to line up 600 nucleotides in the proper order to make genesis DNA. In the great majority of cases the attempt is unsuccessful, but these 'runs', of course, are never seen. Only in that rare case when a run is successful, and life does get started on a planet, is there anyone around to view the film.

The universe, therefore, contains an infinite number of inhabited planets, but the chance that any specific galaxy will contain life is extremely small. Most intelligent races should see no other civilizations in their galaxy; indeed, they should see no others in the entire portion of the universe (including perhaps 10^{22} stars) which they are able to observe with their telescopes. This theoretical prediction is, of course, in complete agreement with our failure to observe extraterrestrials, and with all our other observational evidence.

Conclusion

All of the calculations made above are based on existing theories. No extraordinary assumptions have been made, nor have any unknown effects or processes been postulated. Normally, when theoretical conclusions based on existing theories are in complete accord with the observations the conclusions are readily accepted.

Why, then, are so many people reluctant to believe that N is a low number? I would suggest that this reluctance is primarily a result of wishful thinking: a galaxy teeming with bizarre life forms sounds a lot more interesting than one in which we are alone. But N can be a high number only if $f_{life} \gg 10^{-30}$, and that can be the case only if there exists some abiotic process – as yet totally unknown – which lines up nucleotides in a sequence which is biologically useful. Although we cannot absolutely prove that no such process exists, we should certainly be reluctant to postulate an unknown process when all the observed facts can be explained without it.

References

BROWN, H. (1952). Rare gases and the formation of the Earth's atmosphere. In *The Atmospheres of the Earth and Planets*, 2nd edn, ed. G. P. Kuiper, pp. 258–66. Chicago: University of Chicago Press.

BUDYKO, M. I. (1969). The effect of solar radiation variations on the climate of the Earth. *Tellus*, 21, 611–19.

ELLIS, G. F. R. & BRUNDRIT, G. B. (1979). Life in the infinite universe. *Quart. J. Royal Astron. Soc.*, 20, 37–40.

FEINBERG, G. & SHAPIRO, R. (1980). *Life Beyond Earth*, chapter 6. New York: William Morrow.

GOTT, J. R., GUNN, J. E., SCHRAMM, D. M. & TINSLEY, B. M. (1974). An unbound universe? *Astrophysical Journal*, 194, 543–53.

HART, M. H. (1975). An explanation for the absence of extraterrestrials on Earth. *Quart. J. Royal Astron. Soc.*, 16, 128–35.

HART, M. H. (1978). The evolution of the atmosphere of the Earth. *Icarus*, 33, 23–9.

HART, M. H. (1979). Habitable zones about main sequence stars. *Icarus*, 37, 351–7.

HUANG, S.-S. (1959). Occurrence of life in the universe. *Am. Sci.*, 47, 397–402.

HUANG, S.-S. (1960). Life outside the solar system. *Sci. Am.*, 202, 55–63.

JONES, E. M. (1976). Colonization of the galaxy. *Icarus*, 28, 421–2.

MILLER, S. L. & ORGEL, L. E. (1974). *The Origins of Life on the Earth*, pp. 100–2. Englewood Cliffs, NJ: Prentice-Hall.

PAPAGIANNIS, M. D. (1978). Could we be the only advanced technological civilization in our galaxy? In *Origin of Life*, ed. H. Noda, Tokyo: Center for Academic Publications, Japan Scientific Societies Press.

RASOOL, S. I. & DE BERGH, C. (1970). The runaway greenhouse and the accumulation of CO_2 in the Venus atmosphere. *Nature*, 226, 1037–9.

SELLERS, W. D. (1969). A global climate model based on the energy balance of the Earth–atmosphere system. *J. App. Meteorol.*, 8, 392–400.

SPITZER, L. (1952). The terrestrial atmosphere above 300 km. In *The Atmospheres of the Earth and Planets*, 2nd edn, ed. G. P. Kuiper, pp. 211–47. Chicago: University of Chicago Press.

UREY, H. C. (1951). The origin and development of the Earth and other terrestrial planets. *Geochim. Cosmochim. Acta*, 1, 209–77.

UREY, H. C. (1952). On the early chemical history of the Earth and the origin of life. *Proc. Natl. Acad. Sci. USA*, 38, 351–63.

About the Editors
and Contributors

Michael H. Hart did his undergraduate work at Cornell University and later obtained a PhD in astronomy from Princeton. He also has graduate degrees in physics, in law and in computer science. He has written numerous articles on astronomy for professional journals and is also the author of a book on history: *The 100: A Ranking of the Most Influential Persons in History*. Dr Hart has worked at Hale Observatories in California, at the National Center for Atmospheric Research in Colorado, and at NASA's Goddard Space Flight Center in Maryland. At present, Dr Hart teaches astronomy and history of science at Anne Arundel Community College in Maryland. He is married and has two children.

Benjamin Zuckerman is a professor in the Physics & Astronomy Department at the University of California, Los Angeles. When not ruminating about intelligent life in the universe, his major scientific interests are the birth and death of stars and the prevalence of planetary systems in our Galaxy. He was a codiscoverer of various molecules in interstellar space including formaldehyde, ethyl alcohol and formic acid. He enjoys hiking in remote areas of Utah and Arizona.

Edward Argyle has a master's degree in nuclear physics and a PhD in astronomy. He was a radioastronomer and spent most of his career at the Dominion Radio Astrophysical Observatory in Penticton, British Columbia, where he was a senior research officer until his retirement.

Ronald N. Bracewell is Lewis M. Terman Professor of Electrical Engineering emeritus at Stanford University. A native of Australia, he has degrees in mathematics, physics and electrical engineering. Dr Bracewell is the author of *The Galactic Club: Intelligent Life in Outer Space*, and has also written books on radio-astronomy and applied mathematics.

Ian Crawford is an astronomer specializing in high-resolution optical spectroscopy of the interstellar medium. He holds BSc and PhD degrees in astronomy from the University of London, and an MSc in geophysics and planetary physics from the University of Newcastle upon Tyne. He is deeply interested in the future of humanity in space.

Jared M. Diamond, an evolutionary biologist and physiologist, is Professor of Physiology at the University of California Medical School in Los Angeles. He

226

is a member of the National Academy of Sciences and a regular contributor to *Discover* and *Natural History* magazines. His prize-winning book *The Third Chimpanzee* is a popular account of the evolution of supposedly unique human characteristics from animal precursors. It includes a discussion of human characteristics that may account for the apparent absence of intelligent extraterrestrials.

Freeman Dyson has for many years been a Professor of Physics at the Institute for Advanced Study at Princeton, NJ. He has written for the general public, including *Disturbing the Universe*, *Origins of Life* and *From Eros to Gaia*, besides various journal articles concerning extraterrestrials.

Gerald Feinberg, deceased, was Professor of Physics at Columbia University and served as chairman of the department. He wrote numerous articles for scientific journals and several books for the general reader, which reflected a variety of interests. They included: *The Promethieus Project, Mankind's Search for Long-Range Goals* (1969), *What is the World Made Of?* (1977), *Consequences of Growth* (1977), *Life Beyond Earth: The Intelligent Earthling's Guide to Life in the Universe* (with Robert Shapiro) (1980) and *Solid Clues* (1985).

J. Richard Gott, III is a professor at Princeton University in the Department of Astrophysical Sciences. Before coming to Princeton he did research at the California Institute of Technology and at Cambridge University in England. Professor Gott's research specialities are cosmology and general relativity, but he has also written articles in many other branches of astronomy and astrophysics. In 1975 he received the Trumpler award of the Astronomical Society of the Pacific, and in 1977 was named an Alfred P. Sloan Fellow. He is also chairman of the judges for the Westinghouse Science Talent Search.

Eric M. Jones is a laboratory fellow at the Los Alamos National Laboratory. He has degrees in astronomy from the California Institute of Technology and the University of Wisconsin. He conducts research on nuclear test treaty verification issues and on space development. With anthropologist Ben R. Finney, he edited the book *Interstellar Migration and the Human Experience* and, with the astronauts, is currently compiling the *Apollo Lunar Surface Journal*, a set of annotated transcripts of the Apollo missions.

Gerald F. Joyce is an associate professor in the Departments of Chemistry and Molecular Biology at the Scripps Research Institute. After receiving his BA from the University of Chicago in 1978, he entered the MD/PhD program at the University of California San Diego. His thesis research was carried out under Leslie Orgel at the Salk Institute and concerned nucleic acid chemistry and the template-directed synthesis of RNA. After completing his predoctoral studies, Dr Joyce undertook a medical internship at Mercy Hospital in San Diego and obtained his medical license. He then returned to the laboratory to pursue postdoctoral research with Tan Inoue at the Salk Institute, studying the biochemistry of RNA enzymes. Dr Joyce joined the faculty of the Scripps Research Institute in 1989 and was promoted to associate professor in 1992. His current research involves development of directed evolution techniques and application of these techniques to the study and design of RNA enzymes. He also has a longstanding interest in the origins of life and the role of RNA in the early history of life on Earth.

Jonathan I. Lunine is Associate Professor of Planetary Sciences and Theoretical Astrophysics at the University of Arizona. He obtained his PhD from Caltech in 1985 and taught briefly at UCLA in a visiting position before joining the Arizona faculty. Professor Lunine is the author of roughly 70 papers concerning the origin of the solar system, evolution of planetary surfaces and atmospheres and related theoretical and laboratory studies. He is coeditor of the book *Protostars and Planets III*, and is an Interdisciplinary Scientist on the Cassini mission to Saturn and Titan. He won the 1988 Harold C. Urey Prize of the AAS's Division for Planetary Sciences and one of six 1990 Zeldovich Awards from COSPAR. He currently chairs NASA's Solar System Exploration Subcommittee, which advises the agency on its planetary programs.

Ernst Mayr is Professor of Zoology emeritus at Harvard University. An evolutionist, historian, and philosopher of science, Dr Mayr has been awarded the Balzan Prize, the National Medal of Science and other honors.

Rafael Navarro-González is Research Professor of Chemistry at the Institute of Nuclear Sciences of the National Autonomous University of Mexico. His main research interest is the origin of life, in particular, the synthesis of biomolecules in planetary atmospheres by lightning, and molecular replication. He has been recognized as a Young National Researcher by the Mexican Department of Education since 1990.

James Oberg works for Rockwell as a contractor to NASA in mission control Houston, specializing in orbital rendezvous. He writes regularly for *OMNI* magazine, and has contributed articles to *New Scientist, Star and Sky, Space World, The Skeptical Inquirer, Astronomy* and many other magazines. His books include: *New Earths*, which discusses the possibilities of planetary engineering, *Red Star in Orbit*, a survey for the layman of the Soviet manned space program, and *Uncovering Soviet Disasters*, investigations into Soviet 'official secrets'.

Michael D. Papagiannis was born and raised in Athens, Greece, where he received his master's degree in chemical engineering. He came to the United States as a young man, where he received his MS in physics from the University of Virginia, and his PhD in astronomy from Harvard. He joined the faculty of Boston University, where he was also the chairman of the new Astronomy Department for 14 years. His research specialities are in space physics, radioastronomy and bioastronomy (the search for extraterrestrial intelligence), having served also as the first President of IAU Commission 51-Bioastronomy from 1982 to 1985, and as the secretary of IAU Commission 51 and the editor of its *Bioastronomy News* for 9 years. He is a corresponding member of the Academy of Athens; a full member of the International Academy of Astronautics; member of executive committee of the corporation operating Haystack Radio Obs. 1970–84; trustee of Hellenic Coll.; fellow AAAS, member AAS, AGU, IAU; author *Space Physics and Space Astronomy* (1972 & 1978), editor: *8th Texas Symposium of Relativistic Astrophysics*, 1977, *Strategies for the Search for Life in the Universe*, 1980, *The Search for Extraterrestrial Life: Recent Developments* (IAU Symposium No 112) 1985.

Cyril Ponnamperuma, deceased, was Professor of Chemistry and Director of the Laboratory of Chemical Evolution at the University of Maryland-College Park. His main research interests were the study of the origin of life and the possibility of life beyond the Earth. He was a principal investigator in the Apollo program and was associated with many of NASA's planetary missions. In 1980 he received the first A.I. Oparin award presented by the International Society for the Study of the Origin of Life for 'the best sustained research program' in the origin of life. In 1993, the Russian Academy of Creative Arts awarded him the Harold Urey prize and the academy award for his contribution to the study of chemical evolution.

Robert Shapiro is Professor of Chemistry at New York University. His research interests include the mechanisms by which environmental chemicals cause cancer, and the origin of life. He is the author of about 90 scientific articles and has written three books: *The Human Blueprint* (1991), *Origins, A Skeptic's Guide to the Origin of Life on Earth* (1986) and (with Gerald Feinberg) *Life Beyond Earth: The Intelligent Earthling's Guide to Life in the Universe* (1980).

Robert Sheaffer, a leading skeptical investigator of UFOs, is a computer software specialist in the Silicon Valley, as well as a freelance writer. A fellow of the Committee for the Scientific Investigation of Claims of the Paranormal (CSICOP), as well as a founding member of the Bay Area Skeptics, he actively pursues a lifelong interest in astronomy and the question of life on other worlds. He is the author of *The UFO Verdict*, as well as *Resentment Against Achievement* and *The Making of the Messiah*.

Clifford E. Singer received his PhD in biochemistry at Berkeley in 1971 and then spent 5 years at Queen Mary College in London and 7 years at Princeton. He is now Professor of Nuclear Engineering at the University of Illinois at Urbana-Champaign. He has published research on molecular biology, genetic evolution, solar and lunar physics, plasma physics, extraterrestrial resources, propulsion, nuclear proliferation and controlled fusion.

Jill Tarter received her PhD in astronomy from the University of California at Berkeley, where her major field of study was theoretical high energy astrophysics. As a principal investigator for the non-profit SETI Institute in Mountain View,

CA, Dr Tarter served as project scientist for NASA's High Resolution Microwave Survey (HRMS), formerly known as SETI. As such, she had the opportunity to conduct and plan for thorough observations of the sky through a set of narrowband and pulse-sensitive filters never before systematically employed by astronomers. On 1 October 1993, Congress terminated funds for HRMS in the FY94 Appropriation Bill for NASA. Dr Tarter is currently working with the SETI Institute to secure private funding in order to continue the Targeted Search element of HRMS. Dr Tarter has received several awards for her work, including the Lifetime Achievement Award from Women in Aerospace and two Public Service Medals from NASA. She has authored numerous papers and presents lectures at scientific symposia and colloquia throughout the world.

Virginia Trimble is a native Californian, having received degrees from UCLA (BA 1964), Caltech (MS 1965; PhD 1968), and Cambridge University (MA 1969). She currently shares appointments at the University of California, Irvine, and the University of Maryland with her husband physicist Joseph Weber. She has published papers on the subjects of structure and evolution of stars and galaxies, cosmology, and history and sociology of astronomy and currently serves as editor of *Comments on Astrophysics* and associate editor of the *Astrophysical Journal*.

Sebastian von Hoerner was a scientist at the US National Radio Astronomy Observatory (NRAO). He was born in Germany and studied theoretical physics at Gottingen. He worked at the Max-Planck-Institut in Gottingen and the Astronomisches Recheninstitut in Heidelberg where he investigated diverse problems in turbulence, shock fronts, astrophysical hydrodynamics, stellar dynamics and stellar evolution. Dr von Hoerner's work at the NRAO, for which he now works as a consultant, has concerned radioastronomy, cosmology, life in space, and the design and improvement of radio telescopes. He has published two papers on musical scales and enjoys sailing and hiking.

Author Index

231

Drake, F. & Sobel, D. (1992) 32, 173, 182
Drell, S. D., Foley, H. M. & Ruderman, M. A.
    (1965) 47
Driggers, G. W. (1979) 73
Duncan, M., Quinn, T. & Tremaine, S. (1988)
    198
Dyson, F. J.
    (1960) 10, 180
    (1968) 32, 45, 75, 77–8
    (1979) 95–100, 176

Edmunds, M. G. & Terlevich, R. J. (1992)
    184–8
Ellis, G. & Brundrit, B. (1979) 224

Feinberg, G. & Shapiro, R. (1980) 105, 109,
    220
Ferris, J. P. & Hagan Jr., W. J. (1984) 114
Ferris, J. P. *et al.* (1978) 113
Finney, B. R. & Jones, E. M.
    (1984) 100
    (1985) 50, 93
Fishback, J. F. (1969) 64–5
Flores, J. & Ponnamperuma, C. (1972) 115
Ford, K. W. (1959) 46
Forward, R. I.
    (1962) 56
    (1982) 55
    (1984) 56–7, 59
    (1985) 45
    (1986) 59
Fox, S. W. & Dose, K. (1972) 125
Freeman, A. & Millar, T. J. (1983) 166
Freitas, R. A. (1985) 10
Freitas, R. A. & Valdes, F. (1980) 10
Friebele, D. *et al.* (1981) 117
Friebele, D., Shimoyama, A. & Ponnamperuma,
    C. (1980) 117–18
Frogel, J. A. (1988) 188
Fukuyama, F. (1989) 50

Gabel, N. W. & Ponnamperuma, C. (1967) 112
Gallup, G. (1978) 20
Garrison, W. M. *et al.* (1951) 110
Gatewood, G. D. (1987) 207
Gil, J. A. & Jessner, A. (1993) 210
Gilligan, E. S. (1975) 72
Gilmore, G., Wyser, F. G. & Kuijken, K. (1989)
    186
Goldman, T. (1988) 56
Goldreich, P. & Ward, W. R. (1973) 200
Goldsmith, D. & Owen, T. (1980) 42

Gott, J. R.
    (1980) 175
    (1982) 173
    (1993) 181–2
Gott, J. R. *et al.* (1974) 223
Gott, J. R. & Turner, E. U. (1988) 174
Gould, A. & Loeb, A. (1992) 210
Greenberg, J. M. (1989) 121–2
Grey, J. G (1977) 71
Groth, W. & Suess, H. (1978) 110
Guerrier-Takada, C. *et al.* (1983) 141
Gulick, A. (1955) 117
Guth, A. H. (1981) 176

Haldane, J. B. S. (1929) 108
Hands, J. (1985) 61
Hanel, R. (1979) 121
Harris, M. J. (1990) 10
Hart, M. H.
    (1975) 32, 66, 70, 92, 181, 219
    (1978) 216
    (1979) 165, 217
    (1982) 176
Hart, M. H. & Zuckerman, B. (1982) 40
Hasegawa, M. & Yano, T. (1975) 134
Heidmann, J. & Klein, M. J. (1991) 32
Henderson, L. J. (1913) 167
Hendry, A.
    (1978) 21
    (1979) 21, 24
Henry, R. C. B. (1993) 184
Herbst, E. (1990) 121
Hill, B. (1978) 22
Holland, H. D. (1962) 109
Holmes, D. L. (1984) 61
Horowitz, N. H. (1976) 166–7
Horowitz, P. & Sagan, C. (1993) 178
Hoyle, F. & Wickrama, S. E. (1977) 166
Huang, S. S.
    (1959) 217
    (1960) 217
Hubbard, W. B. (1977) 205
Hulshof, J. & Ponnamperuma, C. (1976)
    115
Hynek, J.
    (1976) 25
    (1977) 26

Iben, I. (1984) 41
Irvine, W. M., Ohishi, M. & Kaifu, N. (1991)
    121
Isaacman, R. & Sagan, C. (1977) 210

Newman, W. I. (1983) 96
Newman, W. I. & Sagan, C. (1981) 93, 96, 99
Nielson, H. B. (1989) 181
Nisbet, E. G. (1985) 120
Norem, P. C. (1969) 45

Oberg, J. (1981) 89
Oliver, B. M. (1977) 178
Oliver, B. M. & Billingham, J. (1971) 70
O'Leary, B. (1978) 73
O'Neill, G. K.
  (1974) 32, 73, 94
  (1977) 32
O'Neill, G. K. & O'Leary, B. (1977) 71–2
O'Neill, G. K. & Snow, W. R. (1979) 76
Oparin, A. (1924) 108
Oró, J.
  (1960) 113
  (1961a) 121
  (1961b) 113

Papagiannis, M. D.
  (1977) 33
  (1978) 219
  (1980) 40
  (1983) 10
  (1984–93) 104
Paprotny, Z. Lehman, J. & Prytz, J.
  (1984) 50
  (1986) 50
  (1987) 50
Parkinson, R. C.
  (1974) 66
  (1978) 52
Peimbert, M. (1986) 187
Penzias, A. A. & Wilson, R. W. (1965) 174, 176, 180
Phillips, J. A., Thorsett, S. E. & Kulkarni, S. R.
  (1993) 211
Podolak, M., Hubbard, W. B. & Pollack, J. B.
  (1993) 201
Podsiadlowski, P., Pringle, J. E. & Rees, M. J.
  (1991) 203
Ponnamperuma, C.
  (1972) 119, 121
  (1976) 121
  (1978) 115
Ponnamperuma, C. et al.
  (1963) 113
  (1969) 110
Ponnamperuma, C., Honda, Y. &
  Navarro-González, R.

  (1990) 118
  (1992) 109, 121
Ponnamperuma, C. & Mack, R. (1965) 115
Ponnamperuma, C. & Peterson, E. (1965) 115
Ponnamperuma, C., Suimoyama, A. & Friebele,
  E. (1982) 117–18
Potter, J. (1965) 99
Powell, C. (1975) 53
Purcell, E.
  (1960) 70
  (1963) 3

Rana, N. C. (1991) 186, 188
Rasleigh, S. C. & Marshall, R. A. (1978) 76
Rasool, S. I. & De Bergh, C. (1970) 216
Raulin, F. (1990) 112
Redding, J. I. (1967) 56
Regis, F. (1985) 61
Reid, C. & Orgel, L. F. (1967) 112
Robinson, B. J. (1976) 125
Ruderman, M. (1974) 170
Ruppe, H. O. (1966) 76

Sagan, C.
  (1963) 62
  (1973) 33
Sampson, A. (1977) 33
Sanchez, R. A., Ferris, J. P. & Orgel, L. F.
  (1967) 113
Sandage, A. & Tamman, C. (1975) 174
Sänger, E. (1957) 62
Saumon, D. et al. (1992) 205
Saxinger, C. & Ponnamperuma, C. (1974) 120
Schidlowski, M. (1988) 149
Schidlowski, M., Hayes, M. & Kaplan, I. R.
  (1983) 149
Schimpl, A., Lemmon, R. M. & Calvin, M.
  (1965) 115
Schopf, J. W. (1993) 148
Schopf, J. W. & Packer, B. M. (1987) 148
Schrödinger, E. (1956) 169
Schwartz, A. W. (1983) 113
Schwartz, A. W. & Goverde, M. (1982) 113
Schwartz, A. W. & Orgel, L. E. (1985) 113
Sellers, W. D. (1969) 217
Senaratne, S. M. (1986) 118–19
Sentinel (1978) 22
Shapiro, R. (1988) 112
Sharpton, V. L. et al.
  (1992) 145
  (1993) 145
Sheaffer, R. (1981) 22, 23

Subject Index